Building
A
Tesla Turbine

AF271395

Written & Illustrated by
Vincent R. Gingery

Published by
David J. Gingery Publishing LLC
P.O. Box 318
Rogersville, MO 65742

Web: http://www.gingerybooks.com
Email: gingery@gingerybooks.com

Printed in the U.S.A.

ISBN 1-878087-29-0

Table Of Contents

Note: The original Tesla patent is included on page 17 of this booklet for informational purposes only. No effort has been made to construct a turbine as an exact replica to the original patent. The intention has been to supply a project that would display the amazing principles of the Tesla Turbine yet be fairly simple to construct.

Acknowledgments

No claim is made as to the originality of the turbine design presented here as it has been a project influenced by information gathered from a variety of sources including, but not limited to, the original Tesla patent, The highly recommended booklet 'The Tesla Disc Turbine' by W.M.J. Cairns, a 1965 Popular Mechanics article titled 'Make A Model Tesla Turbine' by Walter E. Burton and the information gathered on the life of Nikola Tesla was found in the excellent book 'The life And Times Of Nikola Tesla', by Marc J. Seifer.

Introduction

The purpose of this booklet is to show you how to build a Tesla Turbine. It is a fairly easy project consisting of a simple assembly of eighteen 2-3/4" diameter flat discs separated by washers and mounted on a shaft held between two high speed bearings and housed within a round cylinder.

The turbine operates on the principle that all fluids have two prominent properties; adhesion and viscosity. To better understand the principle and what it means for our turbine, it is helpful to know that air, steam and the various gasses are classified as fluids just like water is. So when air or steam enters the turbine under pressure, the properties of adhesion and viscosity come into play and the rapid fluid motion essentially grabs the flat surface of the disks forcing them to rotate along with it as it moves ever rapidly to the center of the turbine. This force of movement on the flat disks in turn develops horsepower as well as very high r.p.m..

An overview of the turbine operation is as follows. . .

A pressurized jet of air enters the manifold as shown in figure 2 and makes its way through to the outer edges of the flat disk assembly where it is forced into the turbine by way of a nozzle. It follows a rapid spiral path around and around and as it does, it drags the disks along with it. Eventually, the air works its way to the center of

Figure 1. The turbine complete.

the turbine and exits through the exhaust ports. The direction the turbine rotates is determined by valves located at either side. Looking at figure 2, note that if valve "A" is open and valve "B" is closed, the turbine will rotate clockwise. Likewise, if valve "B" is open and valve "A" is closed, the turbine will rotate counterclockwise.

The turbine as detailed will rotate in excess of 5000 r.p.m. at a constant air pressure delivery of 80 pounds per square inch. Higher delivery pressures will increase the rpm. But be advised, when exposed to such high rotational speeds inertia stress becomes a concern. The turbine as detailed is constructed of type 304 stainless steel which is a stronger material than typical mild steel which does give it the ability to better withstand the stress of high speed rotation. Even so, care should be exercised, particularly at speeds in excess of 5000 rpm. As a precaution I would advise those who intend on operating this turbine, es-

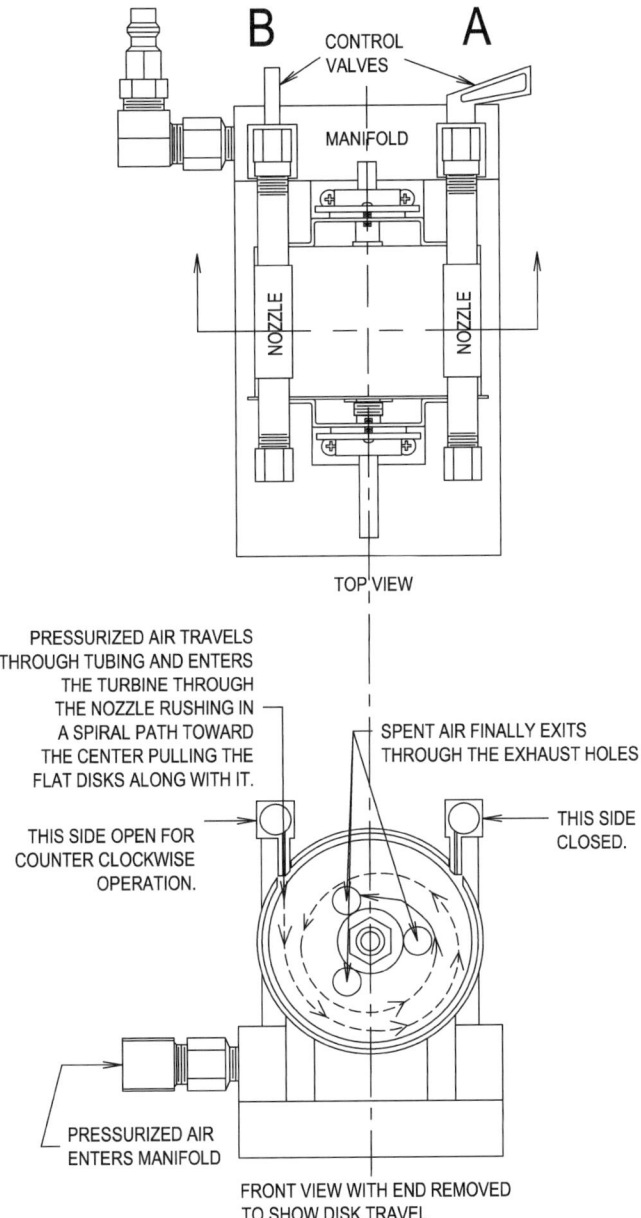

B **A**

CONTROL VALVES

MANIFOLD

NOZZLE   NOZZLE

TOP VIEW

PRESSURIZED AIR TRAVELS
THROUGH TUBING AND ENTERS
THE TURBINE THROUGH
THE NOZZLE RUSHING IN
A SPIRAL PATH TOWARD
THE CENTER PULLING THE
FLAT DISKS ALONG WITH IT.

SPENT AIR FINALLY EXITS
THROUGH THE EXHAUST HOLES

THIS SIDE OPEN FOR
COUNTER CLOCKWISE
OPERATION.

THIS SIDE
CLOSED.

PRESSURIZED AIR
ENTERS MANIFOLD

FRONT VIEW WITH END REMOVED
TO SHOW DISK TRAVEL

Figure 2. The turbine, as detailed here, is set to rotate in a counter clockwise direction. All that is required to switch over to a clockwise rotation is to open the closed valve on the right and close the open valve on the left.

pecially at higher speeds, to devise a safety guard and place it between the turbine and those in close proximity to it.

The turbine project is well suited for those with miniature lathes. The largest turning requirement is the 3-3/16" diameter removable end cap. Besides a lathe, a drill press and/or milling machine and a bandsaw or hacksaw will be required. As can be expected with a project such as this, an assortment of hand tools will also be needed. Such tools as a pair of right hand (red handle) metal cutting snips, hacksaw or bandsaw, a small pipe wrench as well as an assortment of wrenches & screwdrivers as would commonly be found in the average workshop. A set of drill bits, a 4-40 tap, an 8-32 tap, a 1/8-27 N.P.T. tap, a 3/8-24 thread die and a .376 chucking reamer are also required. You will also need a 1/4" & a 1/2" collet to be used in the lathe for turning the rotor shaft.

The turbine itself is constructed entirely of type 304 stainless steel. The rotor shaft is machined from 1/2" diameter s.s. round rod, the disks are cut from 20 gage s.s. sheet metal and the housing (stator) is cut to length from 3" diameter .065" wall s.s. pipe. The pressure line is assembled from an assortment of brass fittings and the end support

brackets are made of brass as well.

Though this project is meant to produce a demonstration model, it does have practical value. Similar turbines have been used to power generators for the purpose of producing electricity. Of course using air pressure from an air compressor would not be economical for such a purpose, but certainly steam pressure would be. However generating electricity and building steam boilers are subjects in themselves and beyond the scope of this book. There is plenty of room for experimentation and it is my hope this project is beneficial in that way.

Remember to be safe. There are hazards lurking in all that we do. You will be working with metal and that includes cutting, forming, grinding and welding all of which pose an injury threat. Wear safety glasses, use common sense and consider each step and procedure carefully before you begin. Remember to safeguard all those who may be in close proximity to you as well. The turbine you are getting ready to build has the potential to develop very high rpm and that in itself poses some danger and you are advised to erect a safety guard between the turbine and those tending it and those who might be looking on as spectators.

Material List

The materials listed below are fairly easy to find. The stainless steel, steel, and brass items can be purchased from steel suppliers. The plumbing fittings can be purchased at any good hardware store or home improvement center and sources for the bearings are listed in the text on page 34.

13" x 20" sheet of 20 gage type 304 s.s. (stainless steel) sheet metal. For rotor disks and end caps. See figures 4, 15 & 17.

2 pieces of 3" x 3" x 1/4" c.r.s. (cold roll steel) flat bar. Jig for drilling rotor disks. See figure 5.

Two 1/4-20 x 1/2" bolts for assembling the above mentioned jig.

1 piece of 5-1/8" long 1/2" diameter type 304 stainless steel round rod. Rotor disk shaft. See figure 9.

Nineteen 3/8" stainless steel flat washers. Rotor disk spacers. See figure 10 & 11.

One 3/8-24 stainless steel nut. For securing rotor disks on shaft. See figure 10.

One 3/8" diameter x 3" long c.r.s round rod. Alignment shaft. See figure 12.

One 2-1/4" long piece of 3" diameter, .065 wall type 304 stainless steel pipe. Stator. See figure 14.

1/4" x 9-1/64" piece of 16 gage brass. Flange ring. See figure 16.

Three 4-40 x 1/4" machine screws. For securing the removable stator end. See figure 18.

One piece of 6" x 3-1/2", 20 gage brass sheet metal. For the two end support brackets. See figure 19.

One 5" long piece of 3/8" diameter c.r.s. round rod. Alignment shaft. See figure 21.

Eight 4-40 x 3/16" brass machine screws. For securing the end support brackets. See figure 21.

2 high speed bearings, .250 I.D. X .750 O.D. X .2812 wide. Sold under part #'s R4A and R4AZZ. See figure 23 and text on page 34 for purchase information.

Material List Continued

One 3" long piece of 1" diameter brass round rod. For making the 2 bearing caps. See figure 24.

One piece of 3" x 1-1/2", 16 gage brass. For making the 2 bearing retainers. See figure 25.

One 3" length of 1" diameter c.r.s. round rod. Arbor for making bearing retainers. See figure 26.

One 1/4-20 x 1/2 bolt with 1/4" flat washer. For arbor mentioned above.

One 1/4" I.D. x .355 O.D. bronze bushing. Will make two spacer bushings. See figure 30.

One 3/4" x 3-1/2" x 6" wooden block. Turbine base. See figure 32.

One 1" x 1" x 3-5/8" length of aluminum bar stock. Manifold. See figure 33.

Two 8-32 x 1" machine screws and two #10 flat washers. For securing manifold to base. See figure 34.

2 pieces of 1/2" x 1" x 1-5/8" aluminum bar stock. For nozzles. See figure 35.

Two 4-40 x 1/4" machine screws with nuts. For securing nozzles mentioned above.

Thread tape or pipe dope for sealing pipe connections.

Two 1/8" FPT brass caps. See figure 38.

Two 4" x 1/8" brass pipe nipples. See figure 38.

Two 1/8 FIP to MIP 90 brass street elbow. See figure 38.

Two 1/8 x 1/8 FPT brass valves. See figure 38.

Two 1/8 x 1-1/2" brass nipples. See figure 38.

Two 1/4 FPT x 1/8 MPT brass adapters. See figure 38.

Two 1/4 FPT to MPT 90 street elbow. See figure 38.

One air coupler plug. See figure 38.

Four #6 x 1/2" wood screws. For securing turbine to base. See figure 43.

Who Was Nikola Tesla?

I receive a number of inquiries on a day to day basis as to what I'm working on at any given moment. Having just begun the Tesla turbine project my first replies were that I was constructing a Tesla turbine. I was amazed to find that most people had never even heard of Nikola Tesla let alone a Tesla turbine.

I wondered. . . How can this be? His inventions, patents and discoveries have molded our modern age yet no one seems to have heard of him. Among his foremost achievements was the discovery of the rotating magnetic field which is the basis for practically all alternating current machinery in existence. Other inventions include the electrical power distribution system we take for granted today, fluorescent and neon lighting, wireless communication, remote control and robotics to name but a few. All told, this profound and amazing man held more than 100 U.S. patents.

Nikola Tesla was born in 1857 in Smiljan, Croatia (Yugoslavia). His family was of Serbian origin and his father was an Orthodox priest who intended his son to follow in his footsteps. His mother was unschooled but extremely intelligent.

Some have referred to Tesla as a dreamer with a poetic touch who added to these qualities self-discipline and a desire for perfection. Such dissertations as 'A Fairy Tale Of Electricity'* written in 1915 give fine insight to the poetic man that Tesla was. And his many and amazing accomplishments give evidence of his incredible drive for excellence.

In reality, Tesla was an eccentric driven by compulsions and in later years, a progressive germ phobia. His intense drive and odd personality allowed him only a small circle of friends. Though he admired intellectual and beautiful women, he had no time to become involved. As is often the case with such manic personalities, genius was the result. And Tesla's genius was displayed in his uncanny ability to somehow sense hidden scientific secrets and then employ his inventive talent to prove his ideas. Tesla indeed was a dreamer. But not only a dreamer, he was a doer. And he was not afraid to dream big, and in his own mind, nothing was impossible.

This statement taken from Tesla's autobiography gives insight into the man Tesla was. . .

"The progressive development of man is vitally dependent on invention. It is the most important

---

For a transcript of 'A Fairy Tale of Electricity' as well as other fascinating information go to the Tesla Museum web site in Belgrade Yugoslavia through the following link. . . **http://www.yurope.com/org/tesla/**

product of his creative brain. Its ultimate purpose is the complete mastery of mind over the material world, the harnessing of the forces of nature to human needs. This is the difficult task of the inventor who is often misunderstood and unrewarded. But he finds ample compensation in the pleasing exercise of his powers and in the knowledge of being one of that exceptionally privileged class without whom the race would have long ago perished in the bitter struggle against pitiless elements. Speaking for myself, I have already had more than my full measure of this exquisite enjoyment, so much that for many years my life was little short of continuous rapture".

Tesla trained for an engineering career by attending the Technical University at Graz, Austria and the University of Prague.

In 1882 he went to work in Paris for Continental Edison Company and in Strassburg in 1883, he constructed his first induction motor.

He sailed to America in 1884 and arrived in New York with just 4 pennies in his pocket. His first job in the U.S. was with Thomas Edison, but the two of them did not get on well.

In May of 1885 George Westinghouse head of the Westinghouse Electric Company bought the patent rights to Tesla's polyphase system of alternating current dynamos, transformers, and motors. This began a titanic power struggle between Edison's direct-current systems and the Tesla-Westinghouse alternating-current approach. As we know today the A.C. approach was the victor.

Meanwhile Tesla had established his own laboratory and continued experimentation on such things as the carbon button lamp, the power of electrical resonance, and on various types of lighting. He often gave exhibitions in his laboratory, lighting lamps without wires by allowing electricity to flow through his own body to discount the fears circulated by the Edison crowd that alternating current was dangerous and even deadly to use as a power source.

In 1893 Westinghouse used Tesla's system to light the World Columbian Exposition at Chicago. His success with doing so opened the door for the contract to install the first power machinery at Niagara Falls, which bore Tesla's name and patent numbers. The project carried power to Buffalo in 1896.

In 1898 Tesla announced his invention of a boat guided by remote control.

In 1899-1900 Tesla set up a laboratory in Colorado Springs Colorado where he discovered what he referred to as terrestrial stationary waves. With this discovery he was able to prove that the earth could be used as a conductor. During the same period in Colorado, he was able to light 200 lamps without wires from a distance of 25 miles and he created man made lightening producing flashes measuring 135 feet.

Later in 1900 he returned to New York to begin construction of a world broadcasting tower that was to be located on Long Island and financed with $150,000.00 provided by J.P. Morgan. Tesla expected this tower to enable worldwide communication and to furnish facilities for sending pictures, messages, weather warnings and stock reports. Because of a financial panic, labor problems and Morgans withdrawal of support the project failed.

Discouraged, Tesla's work shifted to turbines and other projects. Because of a lack of funds many of his ideas remained in his notebooks which are still of great interest today.

In 1917 Tesla was the recipient of the Edison Medal which was the highest honor the American Institute of Electrical Engineers could bestow. However because of his bitter battles and differences with Edison the medal was likely more of an insult to Tesla than an honor.

Tesla was a very complex individual and the remaining years of his life were spent struggling to find purpose and to develop his many ideas some of which were considered quite bizarre. Lacking in personal funds and financial backing his efforts were frustrated. He had been betrayed by those whom he had trusted. Many of them profited on his inventions without giving due credit or financial return to the inventor. He lived the last few years of his life in the Hotel New Yorker and was known to be especially fond of the pigeons in the area. Concerning a pigeon, Tesla wrote this in his later years. . .

"One night as I was lying in bed in the dark, solving problems as usual, (my beloved pigeon) flew through the open window and stood on my desk. As I looked at her I knew she wanted to tell me she was dying. And then as I got her message, there came a light from her eyes; powerful beams of light. When that pigeon died, something went out of my life. I knew my life's work was finished."

The man who was responsible for much of what we take for granted in our modern age, Nikola Tesla died on January 7, 1943 nearly penniless. He was 86 years old. Close to 2000 people including many dignitaries attended his funeral. The custodian of Alien Property impounded his trunks, which held his papers, diplomas and other honors, as well as his letters and laboratory notes. These were eventually inherited by Tesla's nephew, Sava Kosanovich, and later housed in the Nikola Tesla Museum located in Belgrade Yugoslavia.

At the time of Tesla's death the following editorial appeared in the New York Sun. . .

Mr. Tesla was eighty six years old when he died. He died alone. He was an eccentric, whatever that means. A nonconformist, possibly. At any rate, he would leave his experiments and go for a time to feed the silly and inconsequential pigeons in Herald Square. He delighted in talking nonsense; or was it? Granting that he was a difficult man to deal with, and that sometimes his predictions would affront the ordinary human's intelligence, here, still, was an extraordinary man of genius. He must have been. He was seeing a glimpse into that confused and mysterious frontier which divides the known and the unknown...But today we do know that Tesla, the ostensibly foolish old gentleman at times was trying with superb intelligence to find the answers. His guesses were right so often that he would be frightening. Probably we shall appreciate him better a few million years from now.

Tesla's Flat Disk Turbine Described In His Own Words
From an interview appearing in the New York Herald Tribune on Oct. 15, 1911

"I have accomplished what mechanical engineers have been dreaming about ever since the invention of steam power," replied Dr. Tesla. "That is the perfect rotary engine. It happens that I have also produced an engine which will give at least twenty-five times as much power to a pound of weight as the lightest weight engine of any kind that has yet been produced.

"In doing this I have made use of two properties which have always been known to be possessed by all fluids, but which have not heretofore been utilized. These properties are adhesion and viscosity.

"Put a drop of water on a metal plate. The drop will roll off, but a certain amount of the water will remain on the plate until it evaporates or is removed by some absorptive means. The metal does not absorb any of the water, but the water adheres to it.

"The drop of water may change its shape, but until its particles are separated by some external power it remains intact. This tendency of all fluids to resist molecular separation is viscosity. It is especially noticeable in the heavier oils.

"It is these properties of adhesion and viscosity that cause the "skin friction" that impedes a ship in its progress through the water or an aeroplane in going through the air. All fluids have these qualities—and you must keep in mind that air is a fluid, all gases are fluids, steam is fluid. Every known means of transmitting or developing mechanical power is through a fluid medium.

"Now, suppose we make this metal plate that I have spoken of circular in shape and mount it at its centre on a shaft so that it can be revolved. Apply power to rotate the shaft and what happens? Why, whatever fluid the disk happens to be revolving in is agitated and dragged along in the direction of rotation, because the fluid tends to adhere to the disk and the viscosity causes the motion given to the adhering particles of the fluid to be transmitted to the whole mass. Here, I can show you better than tell you."

Dr. Tesla led the way into an adjoining room. On a desk was a small electric motor and mounted on the shaft were half a dozen flat disks, separated by perhaps a sixteenth of an inch from one another, each disk being less than that in thickness. He turned a switch and the motor began to buzz. A wave of cool air was immediately felt.

"There we have a disk, or rather a series of disks, revolving in a fluid—the air," said the inventor. "You need no proof to tell you that the air is being agitated and propelled violently. If you will hold your hand over the centre of these disks—you see the centres have been cut away—you will feel the suction as air is drawn in to be expelled from the peripheries of the disks.

"Now, suppose these revolving disks were enclosed in an air tight case, so constructed that the air could enter only at one point and be expelled only at another—what would we have?"

"You'd have an air pump," I suggested.

"Exactly—an air pump or blower," said Dr. Tesla.

"There is one now in operation delivering ten thousand cubic feet of air a minute.

"Now, come over here." He stepped across the hall and into another room, where three or four draughtsmen were at work and various mechanical and electrical contrivances were scattered about. At one side of the room was what appeared to be a zinc or aluminum tank, divided into two sections, one above the other, while a pipe that ran along the wall above the upper division of the tank was connected with a little aluminum case about the size and shape of a small alarm clock. A tiny electric motor was attached to a shaft that protruded from one side of the aluminum case. The lower division of the tank was filled with water.

"Inside of this aluminum case are several disks mounted on a shaft and immersed in a fluid, water," said Dr. Tesla. "From this lower tank the water has free access to the case enclosing the disks. This pipe leads from the periphery of the case. I turn the current on, the motor turns the disks and as I open this valve in the pipe the water flows."

He turned the valve and the water certainly did flow. Instantly a stream that would have filled a barrel in a very few minutes began to run out of the pipe into the upper part of the tank and thence into the lower tank.

"This is only a toy," said Dr. Tesla. "There are only half a dozen disks— 'runners,' I call them— each less than three inches in diameter, inside of that case. They are just like the disks you saw on the first motor—no vanes, blades or attachments of any kind. Just perfectly smooth, flat disks revolving in their own planes and pumping water because of the viscosity and adhesion of the fluid. One such pump now in operation, with eight disks, eighteen inches in diameter, pumps four thousand gallons a minute to a height of 360 feet."

We went back into the big, well lighted office. I was beginning to grasp the new Tesla principle.

"Suppose now we reversed the operation," continued the inventor. "You have seen the disks acting as a pump. Suppose we had water, or air under pressure, or steam under pressure, or gas under pressure, and let it run into the case in which the disks are contained—what would happen?"

"The disks would revolve and any machinery attached to the shaft would be operated—you would convert the pump into an engine," I suggested.

"That is exactly what would happen—what does happen," replied Dr. Tesla. "It is an engine that does all that engineers have ever dreamed of an engine doing, and more. Down at the Waterside power station of the New York Edison Company, through their courtesy, I have had a number of such engines in operation. In one of them the disks are only nine inches in diameter and the whole working part is two inches thick. With steam as the propulsive fluid it develops 110-horse power, and could do twice as much."

"You have got what Professor Langley was trying to evolve for his flying machine—an engine that will give a horse power for a pound of weight," I suggested.

"I have got more than that," replied Dr. Tesla. "I have an engine that will give ten horse power to the pound of weight. That is twenty-five times as powerful as the lightest weight engine in use today. The lightest gas engine used on aeroplanes weighs two and one-half pounds to the horse power. With two and one-half pounds of weight I can develop twenty-five horse power."

"That means the solution of the problem of flying," I suggested.

"Yes, and many more," was the reply. "The applications of this principle, both for imparting power to fluids, as in pumps, and for deriving power from fluids, as in turbines, are boundless. It costs almost nothing to make, there is nothing about it to get out of order, it is reversible, simply have two ports for the gas or steam to enter by, one on each side, and let it into one side or the other.

I remembered the bushels of broken blades that were gathered out of the turbine casings of the first turbine equipped steamship to cross the ocean, and realized the importance of this phase of the new engine.

"Then, too," Dr. Tesla went on, "there are no delicate adjustments to be made. The distance between the disks is not a matter of microscopic accuracy and there is no necessity for minute clearances between the disks and the case. All one needs is some disks mounted on a shaft, spaced a little distance apart and cased so that a fluid can enter at one point and go out at another. If the fluid enters at the centre and goes out at the periphery it is a pump. If it enters at the periphery and goes out at the center it is a motor.

"Coupling these engines in series, one can do away with gearing in machinery. Factories can be equipped without shafting. The motor is especially adapted to automobiles, for it will run on gas explosions as well as on steam. The gas or steam can be let into a dozen ports all around the rim of the case if desired. It is possible to run it as a gas engine with a continuous flow of gas, gasoline and air being mixed and the continuous combustion causing expansion and pressure to operate the motor. The expansive power of steam, as well as its propulsive power, can be utilized as in a turbine or a reciprocating engine. By permitting the propelling fluid to move along the lines of least resistance a considerably larger proportion of the available power is utilized.

"As an air compressor it is highly efficient. There is a large engine of this type now in practical operation as an air compressor and giving remarkable service. Refrigeration on a scale hitherto never attempted will be practical, through the use of this engine in compressing air, and the manufacture of liquid air commercially is now entirely feasible.

N. TESLA.

TURBINE.

APPLICATION FILED JAN. 17, 1911.

1,061,206. Patented May 6, 1913.

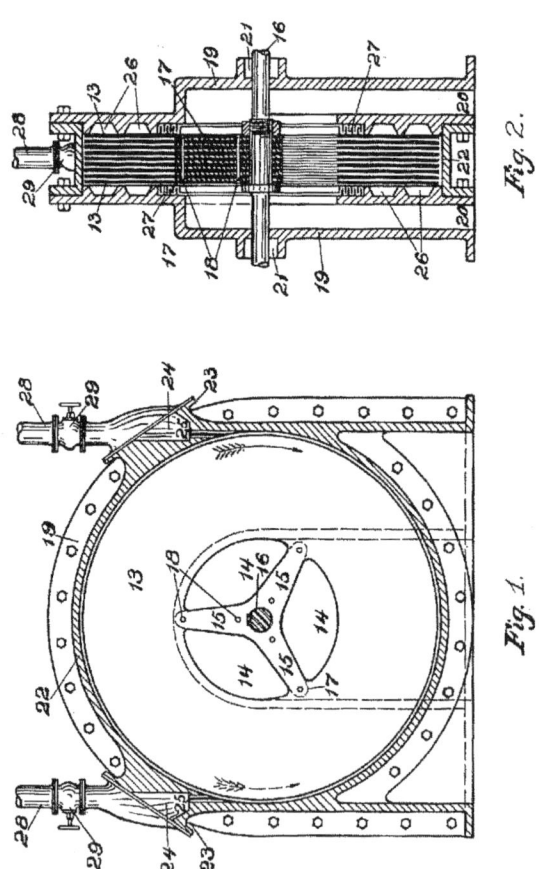

Fig. 2.

Fig. 1.

Witnesses:
R. Diaz Buitrago
Wm. Bohleber

Nikola Tesla, Inventor

By his Attorneys
Kerr Page Cooper & Hayward

UNITED STATES PATENT OFFICE

NIKOLA TESLA, OF NEW YORK, N. Y.

TURBINE

1,061,206.   Specification of Letters Patent.   **Patented May 6, 1913.**

Original application filed October 21, 1909, Serial No. 523,832. Divided and this application filed January 17, 1911.   Serial No. 603,049,

To all whom it may concern:

Be it known that I, Nikola Tesla, a Citizen of the United States, residing at New York, in the county and State of New York, have invented certain new and useful Improvements in Rotary Engines and Turbines, of which the following is a full, clear, and exact description..

In the practical application of mechanical power, based on the use of fluid as the vehicle of energy, it has been demonstrated that, in order to attain the highest economy, the changes in the velocity and direction of movement of the fluid should be as gradual as possible. In the forms of apparatus heretofore devised or proposed, more or less sudden changes, shocks and vibrations are unavoidable. Besides, the employment of the usual devices for imparting to, or deriving energy from a fluid, such as pistons, paddles, vanes and blades, necessarily introduces numerous defects and limitations and adds to the complication, cost of production and maintenance of the machines.

The object of my invention is to overcome these deficiencies and to effect the transmission and transformation of mechanical energy through the agency of fluids in a more perfect manner. And by means simpler and more economical than those heretofore employed. I accomplish this by causing the propelling fluid to move in natural paths or stream lines of least resistance, free from constraint and disturbance such as occasioned by vanes or kindred devices, and to change its velocity and direction of movement by imperceptible degrees, thus avoiding the losses due to sudden variations while the fluid is imparting energy.

It is well known that a fluid possesses, among others, two salient properties, adhesion and viscosity. Owing to these a solid body propelled through such a medium encounters a peculiar impediment known as skin resistance, shock of the fluid against the asperities of the solid substance, the other from internal forces opposing molecular separation. As an inevitable consequence a certain amount of the fluid is dragged along by the moving body. Conversely, if the body be placed in a fluid in motion, for the same reasons, it is impelled in the direction of movement. These 5 effects, in themselves, are of daily observation, but I believe that I am the first to apply them in a practical and economical manner in the propulsion of fluids or in their use as motive agents.

In an application filed by me October 21st, 1909, Serial Number 523,832 of which this case is a division, I have illustrated the principles underlying my discovery as embodied in apparatus designed for the propulsion of fluids. The same principles, however, are capable of embodiment also in that field of mechanical engineering which is concerned in the use of fluids as motive agents, for while in certain respects the operations in the latter case are directly opposite to those met with in the propulsion of fluids, and the means employed may differ in some features, the fundamental laws applicable in the two-cases are the same. In other words, the operation is reversible, for if water or air under pressure be admitted to the opening constituting the outlet of a pump or blower as described, the runner is set in rotation by reason of the peculiar properties of the fluid which, in its movement through the device, imparts its energy thereto.

The present application which is a division of that referred to, is specially intended to describe and claim my discovery above set forth, so far as it bears on the use of fluids as motive agents, as distinguished from the applications of the same to the propulsion or compression of fluids.

In the drawings, therefore, I have illustrated only the form of apparatus designed for the thermo dynamic conversion of energy, a field in which the applications of the principle have the greatest practical value.

Figure 1 is a partial end view, and Fig. 2 a vertical cross-section of a rotary engine or turbine, constructed and adapted to be operated in accordance with the principles of my invention.

The apparatus comprises a runner composed of a plurality of flat rigid disks 13 of suitable diameter, keyed to a shaft 16 and held in position thereon by a threaded nut 11, a shoulder 12, and intermediate washers 17. The disks have openings 14 adjacent to the shaft and spokes 15, which may be substantially straight. For the sake of clearness, but a few disks, with comparatively wide intervening spaces, are illustrated.

The runner is mounted in a casing comprising two end castings 19, which contain the bearings for the shaft 16, indicated but not shown in detail; stuffing boxes 21 and outlets 20. The end castings are united by a central ring 22, which is bored out to a circle of a slightly larger diameter than that of the disks, and has flanged extensions 23, and inlets 24, into which finished ports or nozzles 25 are inserted. Circular grooves 26 and labyrinth packing 27 are provided on the sides of the runner. Supply pipes 28, with valves 29, are connected to the flanged extensions of the central ring, one of the valves being normally closed.

For a more ready and complete understanding of the principle of operation it is of advantage to consider first the actions that take place when the device is used for the propulsion of fluids for which purpose let it be assumed that power is applied to the shaft and the runner set in rotation say in a clockwise direction. Neglecting, for the moment, those features of construction that make for or against the efficiency of the device as a pump, as distinguished from a motor, a fluid,

by reason of its properties of adherence and viscosity, upon entering through the inlets 20, and coming in contact with the disks 13, is taken hold of by the latter and subjected to two forces, one acting tangentially in the direction of rotation, and the other radially outward. The combined effect of these tangential and centrifugal forces is to propel the fluid with continuously increasing velocity in a spiral path until it reaches a suitable peripheral outlet from which it is ejected. This spiral movement, free and undisturbed and essentially dependent on the properties of the fluid, permitting it to adjust itself to natural paths or stream lines and to change its velocity and direction by insensible degrees, is a characteristic and essential feature of this principle of operation.

While traversing the chamber inclosing the runner, the particles of the fluid may complete one or more turns, or but a part of one turn, the path followed being capable of close calculation and graphic representation but fairly accurate estimates of turns can be obtained simply by determining the number of revolutions required to renew the fluid passing through the chamber and multiplying it by the ratio between the mean speed of the fluid and that of the disks. I have found that the quantity of fluid propelled in this manner, is, other conditions being equal, approximately proportionate to the active surface of the runner and to its effective speed. For this reason, the performance of such machines augments at an exceedingly high rate with the increase of their size and speed of revolution.

The dimensions of the device as a whole, and the spacing of the disks in any given machine will be determined by the conditions and requirements of special cases. It may be stated that the intervening distance should be the greater, the larger the diameter of the disks, the longer the spiral path of the fluid and the greater its viscosity. In general, the spacing should be such that the entire mass of the fluid, before leaving the runner, is accelerated to a nearly uniform velocity, not much below that of the periphery of the disks under normal working conditions, and almost equal to it when the outlet is closed and the particles move in concentric circles.

18

Considering now the converse of the above described operation and assuming that fluid under pressure be allowed to pass through the valve at the side of the solid arrow, the runner will be set in rotation in a clockwise direction, the fluid traveling in a spiral path go and with continuously diminishing velocity until it reaches the orifices 14 and 20, through which it is discharged. If the runner be allowed to turn freely, in nearly frictionless bearings, its rim will attain a speed closely approximating the maximum of that of the adjacent fluid and the spiral path of the particles will be comparatively long, consisting of many almost circular turns. If load is put on and the runner slowed down, the motion of the fluid is retarded, the turns are reduced, and the path is shortened.

Owing to a number of causes affecting the performance, it is difficult to frame a precise rule which would be generally applicable, but it may be stated that within certain limits, and other conditions being the same, the torque is directly proportionate to the square of the velocity of the fluid relatively to the runner and to the effective area of the disks and, inversely, to the distance separating them. The machine will, generally, perform its maximum work when the effective speed of the runner is one-half of that of the fluid; but to attain the highest economy, the relative speed or slip, for any given performance should be as small as possible. This condition may be to any desired degree approximated by increasing the active area of and reducing the space between the disks.

When apparatus of the kind described is employed for the transmission of power certain departures from similarity between transmitter and receiver are necessary for securing the best results. It is evident that, when transmitting power from one shaft to another by such machines, any desired ratio between the speeds of rotation may be obtained by a proper selection of the diameters of the disks, or by suitably staging the transmitter, the receiver or both. But it may be pointed out that in one respect, at least, the two machines are essentially different. In the pump, the radial or static pressure, due to

centrifugal force, is added to the tangential or dynamic, thus increasing the effective head and assisting in the expulsion of the fluid. In the motor, on the contrary, the first named pressure, being opposed to that of supply, reduces the effective head and the velocity of radial flow toward the center. Again, in the propelled machine a great torque is always desirable, this calling for an increased number of disks and smaller distance of separation, while in the propelling machine, for numerous economic reasons, the rotary effort should be the smallest and the speed the greatest practicable. Many other considerations, which will naturally suggest themselves, may affect the design and construction, but the preceding is thought to contain all necessary information in this regard.

In order to bring out a distinctive feature, assume, in the first place, that the motive medium is admitted to the disk chamber through a port, that is a channel which it traverses with nearly uniform velocity. In this case, the machine will operate as a rotary engine, the fluid continuously expanding on its tortuous path to the central outlet. The expansion takes place chiefly along the spiral path, for the spread inward is opposed by the centrifugal force due to the velocity of whirl and by the great resistance to radial exhaust. It is to be observed that the resistance to the passage of the fluid between the plates is, approximately, proportionate to the square of the relative speed, which is maximum in the direction toward the center and equal to the full tangential velocity of the fluid. The path of least resistance, necessarily taken in obedience to a universal law of motion is, virtually, also that of least relative velocity. Next, assume that the fluid is admitted to the disk chamber not through a port, but a diverging nozzle, a device converting wholly or in part, the expansive into velocity-energy. The machine will then work rather like a turbine, absorbing the energy of kinetic momentum of the particles as they whirl, with continuously decreasing speed, to the exhaust.

The above description of the operation, I may add, is suggested by experience and observation, and is advanced merely for the purpose of explanation. The undeniable fact is that the machine does operate, both expansively and impulsively. When the expansion in the nozzles is complete, or nearly so, the fluid pressure in the peripheral clearance space is small; as the nozzle is made less divergent and its section enlarged, the pressure rises, finally approximating that of the supply. But the transition from purely impulsive to expansive action may not be continuous throughout, on account of critical states and conditions and comparatively great variations of pressure may be caused by small changes of nozzle velocity. In the preceding it has been assumed that the pressure of supply is constant or continuous, but it will be understood that the operation will be, essentially the same if the pressure be fluctuating or intermittent, as that due to explosions occurring in more or less rapid succession.

A very desirable feature, characteristic of machines constructed and operated in accordance with this invention, is their capability of reversal of rotation. Fig. 1, while illustrative of a special case, may be regarded as typical in this respect. If the right hand valve be shut off and the fluid supplied through the second pipe, the runner is rotated in the direction of the dotted arrow, the operation, and also the performance remaining the same as before the central ring being bored to a circle with this purpose in view. The same result may be obtained in many other ways by specially designed valves, ports or nozzles for reversing the flow, the description of which is omitted here in the interest of simplicity and clearness. For the same reasons but one operative port or nozzle is illustrated which might be adapted to a volute but does not fit best a circular bore. It will be understood that a number of suitable inlets may be provided around the periphery of the runner to improve the action and that the construction of the machine may be modified in many ways.

Still another valuable and probably unique quality of such motors or prime movers may be described. By proper construction and observance of working conditions the centrifugal pressure, opposing the passage of the fluid, may, as already indicated, be made nearly equal to the pressure of supply when the machine is running idle. If the inlet section be large, small changes in the speed of revolution will produce great differences in flow which are further enhanced by the concomitant variations in the length of the spiral path. A self-regulating machine is thus obtained bearing a striking resemblance to a direct-current electric motor in this respect that, with great differences of impressed pressure in a wide open channel the flow of the fluid through the same is prevented by virtue of rotation. Since the centrifugal head increases as the square of the revolutions, or even more rapidly, and with modern high grade steel great peripheral velocities are practicable, it is possible to attain that condition in a single stage machine, more readily if the runner be of large diameter. Obviously this problem is facilitated by compounding, as will be understood by those skilled in the art. Irrespective of its bearing on economy, this tendency which is, to a degree, common to motors of the above description, is of special advantage in the operation of large units, as it affords a safeguard against running away and destruction. Besides these, such a prime mover possesses many other advantages, both constructive and operative. It is simple, light and compact, subject to but little wear, cheap and exceptionally easy to manufacture as small clearances and accurate milling work are not essential to good performance. In operation it is reliable, there being no valves, sliding contacts or troublesome vanes. It is almost free of windage, largely independent of nozzle efficiency and suitable for high as well as for low fluid velocities and speeds of revolution.

It will be understood that the principles of construction and operation above generally set forth, are capable of embodiment in machines of the most widely different forms and adapted for the greatest variety of purposes. In my present specification I have sought to describe and explain only the general and typical applications of the principle which I believe I am the first to realize and turn to useful account.

What I claim is:

1. A machine adapted to be propelled by a fluid consisting in the combination with a casing having inlet and outlet ports at the peripheral and central portions, respectively, of a rotor having plane spaced surfaces between which the fluid may flow in natural spirals and by adhesive and viscous action impart its energy of movement to the rotor, as described.

2. A machine adapted to be propelled by a fluid, comprising a rotor composed of a plurality of plane spaced disks mounted on a shaft and open at or near the same, an inclosing casing with a peripheral inlet or inlets, in the plane of the disks, and an outlet or outlets in its central portion, as described.

3. A rotary engine adapted to be propelled by adhesive and viscous action of a continuously expanding fluid comprising in combination a casing forming a chamber, an inlet or inlets tangential to the periphery of the same, and an outlet or outlets in its central portion, with a rotor composed of spaced disks mounted on a shaft, and open or near the same, as described.

4. A machine adapted to be propelled by fluid, consisting in the combination of a plurality of disks mounted on a shaft and at or near the same, and an inclosing casing with ports or passages of inlet and outlet at the peripheral and central portions, respectively, the disks being spaced to form passages through which the fluid may flow, under the combined influence of radial and tangential forces, in a natural spiral path from the periphery toward the axis of, the disks, and impart its energy of movement to the same by its adhesive and viscous action thereon, as set forth.

5. A machine adapted to be propelled by fluid comprising in combination a plurality of spaced disks rotatably mounted and having plane surfaces, an inclosing casing and ports or passages of inlet and outlet adjacent to the periphery and center of the disks, respectively, as set forth.

6. A machine adapted to be propelled by a fluid comprising in combination a runner composed of a plurality of disks having plane surfaces and mounted at intervals on a central shaft, and formed with openings near their centers, and means for admitting the propelling fluid into the spaces between the disks at the periphery and discharging it at the center of the same, as set forth.

7. A thermo-dynamic converter, comprising in combination a series of rotatably mounted spaced disks with plane surfaces, an inclosing casing, inlet ports it the peripheral portion and outlet ports leading from the central portion of the same, as set forth.

8. A thermo-dynamic converter, comprising in combination a series of rotatably mounted spaced disks with plane surfaces and having openings adjacent to their central portions, an inclosing casing, inlet ports in the peripheral portion, and outlet ports leading from the central portion of the same, as set forth.

In testimony whereof I affix my signature in the presence of two subscribing witnesses.

NIKOLA TESLA.

Witnesses:
M. Lawson Dyer,
Wm. Bohleber.

Building the Rotor

The first items to make are the rotor disks shown in figure 4 and you will need 18 of them. They are each cut from a 13" x 20" sheet of 20 gage type 304 stainless steel. The 3" diameter of each disk is laid out on the sheet with a pair of dividers set for a 1-1/2" radius. After some experimentation, I found the best way to cut the disks was with right angle (red handle) aviation tin snips. Use a pair of sharp snips and you should be able to cut the disks from the sheet with ease. However after cutting out 18 disks your hand could be a bit on the stiff side. Also, watch out for the very sharp edges created as you cut out the disks. A pair of leather gloves will help in this situation, but even so the sharp edges can still get you.

Three 3/8" exhaust holes and one 25/64" center hole are located in each disk. In order to build a balanced turbine, it is critical that these holes be precisely located in rela-

tion to each other. You can achieve this with the aide of the simple clamping jig detailed in the next couple of paragraphs. With the aide of the jig you can secure each disk in a fixed position and drill the required holes so that each disk will be part of a matched set.

The jig consists of two pieces of 3" x 3" x 1/4" c.r.s. flat bar. Two 17/64" holes are drilled in the top half of the jig as shown in figure 5.

Figure 3. Using right hand tin snips to cut the rotor disks from a sheet of 20 gage type 304 stainless steel.

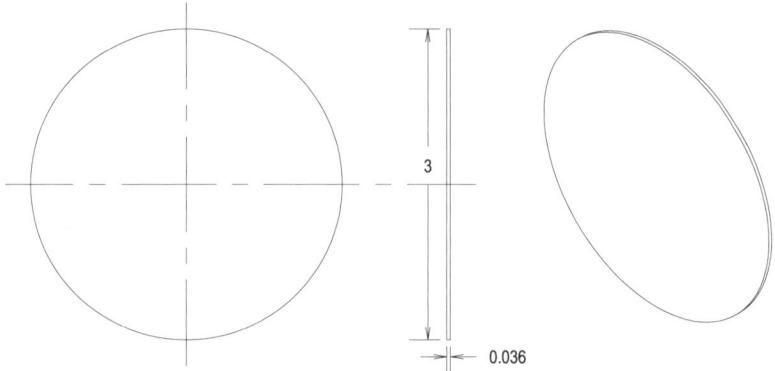

3

0.036

Figure 4. Rotor disks. Cut 18 from a sheet of 20 ga. type 304 stainless steel.

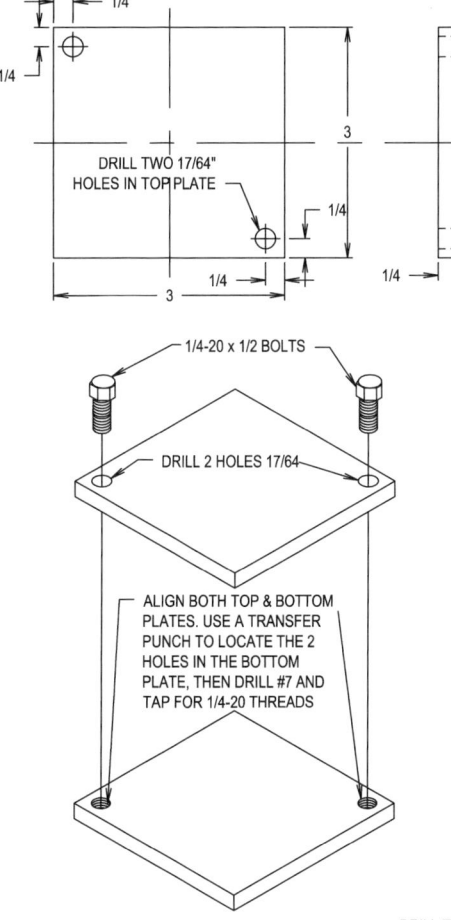

Figure 5. Jig for drilling rotor disks. Made from two pieces of 1/4" x 3" x 3" c.r.s. flat bar.

Then place both the top and bottom halves together and align all four edges. Using a transfer punch inserted through the 17/64" holes in the top half, mark two hole locations in the bottom half. In the marked locations in the bottom half, drill two holes with a #7 drill

and then tap those holes for 1/4-20 thread. Assemble the two halves with two 1/4-20 x 1/2" bolts.

The next step is to layout the location of and drill the three 3/8" holes and the 25/64" center hole in the jig as shown in figure 6. To do that, first find the center of the clamp face by drawing diagonal lines from corner to corner. Mark the center location with a center punch. Then set the dividers at 5/8" apart and scribe a 5/8" radius center circle. Mark the location where the 5/8" circle intersects the center line on the right side of the jig with a center punch. From that marked location and

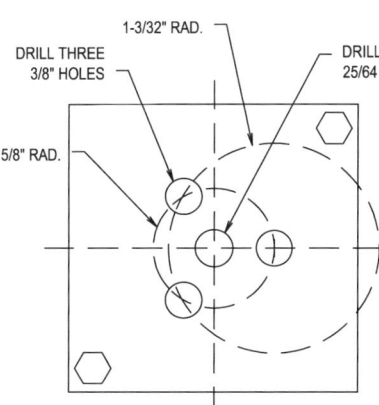

Figure 6. Hole layout for rotor disk jig.

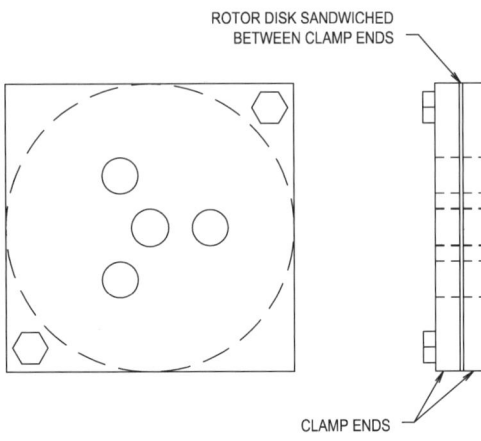

ROTOR DISK SANDWICHED
BETWEEN CLAMP ENDS

CLAMP ENDS

Figure 7. Detail showing rotor disk clamped in jig.

with the dividers set at 1-3/32" apart, scribe a 1-3/32" radius circle. Mark the two points where the 1-3/32" circle passes through the 5/8" radius circle with a center punch. After the layout is complete, drill the 25/64" center hole and the three 3/8" holes at the marked intersection points as shown in the drawing.

The jig is now complete and you are ready to load a rotor disk. Loosen the bolts securing the clamp halves and insert a rotor disk blank between them. Align the edges of the disk with the four sides of the clamp. See figure 7.

Tighten the bolts to secure the disk in the jig. Use a drill press to drill the three 3/8" diameter exhaust holes and the 25/64" center hole. Loosen the clamp bolts and as you remove the drilled disk from the jig, mark the location of one of 3/8" exhaust holes both on the jig for fu-

ture reference and on the rotor disk. A line or arrow pointing from the outer edge of the disk to the hole drawn with a magic marker will serve well for this purpose. See figure 12 & 13.

Repeat the above drilling procedure marking the same corresponding hole on each of the 17 remaining disks. The marked hole location on each disk will be used along with a 3/8" pilot shaft to align the disks as they are mounted on the main shaft a little later. But before the disks can be mounted on shaft we first have to make one, so that's the next order of business.

The rotor shaft is made from a 5-1/8" length of 1/2" dia. stainless steel round rod. Since it is important that the shaft be as straight and

Figure 8. A rotor disk clamped in the jig as the center hole is being drilled.

24

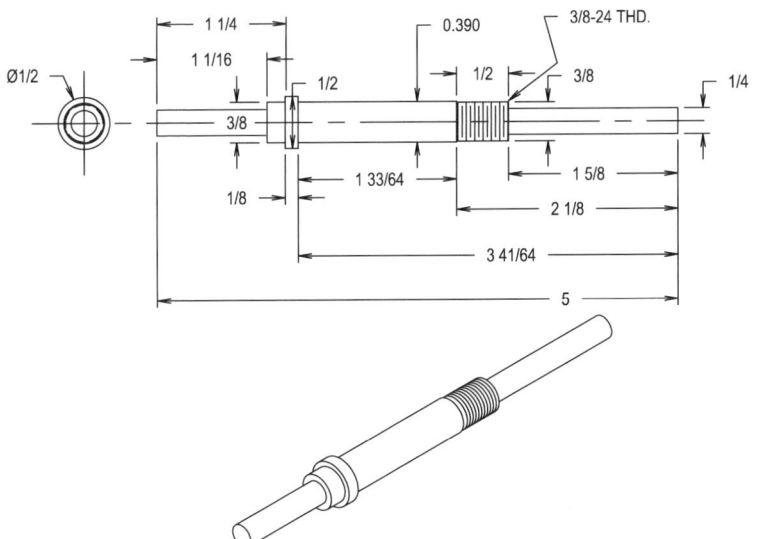

Figure 9. Rotor shaft. Make 1 from a 5-1/8" length of 1/2" dia. s.s. round rod.

balanced as possible, it is best to chuck it in a collet and turn it between centers. To prepare the shaft for turning, chuck one end in a 1/2" collet and face the end off, then countersink to fit a live center. Reverse the shaft in the collet to face off and countersink the opposite end for a live center. Finished length of the shaft to be 5".

With both ends prepared for the live center, adjust one end of the shaft to extend out past the end of the collet at least 3-3/4". Adjust the tail stock forward to secure the live center.

Shaft dimensions are given in figure 9. To make the shaft, begin by reducing the diameter to .390" back 3-41/64" from the end. Next, Reduce the diameter to 3/8" back 2-1/8" from the end. Then reduce the diameter to 1/4" back 1-5/8"

from the end. Finally, thread the 3/8-24 section as shown in the figure. Leave the shaft chucked in the lathe as we will be mounting the rotor disks on it next.

Each rotor disk is aligned on the shaft using the marked hole in each disk as a reference and a 3" length of 3/8" diameter round rod as a pilot shaft. So in order to assemble the turbine you will need to cut a 3/8" diameter x 3" long piece of c.r.s. round rod for the pilot shaft. But before the 3/8" diameter pilot shaft will fit in the marked holes, they must first be enlarged to .376". To accomplish this task, use a .376" chucking reamer and ream the marked hole location in each of the 18 rotor disks.

As you assemble the turbine, each rotor disk will be separated by a 3/8" stainless steel flat washer.

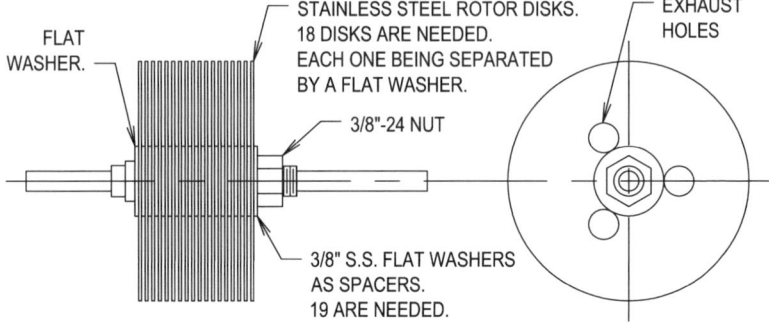

STAINLESS STEEL ROTOR DISKS. 18 DISKS ARE NEEDED. EACH ONE BEING SEPARATED BY A FLAT WASHER.

FLAT WASHER.

EXHAUST HOLES

3/8"-24 NUT

3/8" S.S. FLAT WASHERS AS SPACERS. 19 ARE NEEDED.

Figure 10. Mount the 18 rotor disks on the shaft separating each with a 3/8" stainless steel flat washer.

You will need 19 such washers. These should be readily available from most hardware or home improvement stores. Mine came in plastic bags containing 5 each. The drawing in figure 11 gives the washer dimensions.

Now, with the shaft remaining secured in the lathe collet from the un-threaded end, begin assembling the turbine, first by sliding a spacer washer on the shaft then a rotor disk then a spacer washer. Repeat the sequence until all disks are on the shaft, then finish with a spacer washer at the finished end. Loosely thread a 3/8"-24 nut on the shaft.

Ensure the alignment remains correct, then tighten the nut with a wrench to secure the disks on the shaft. See figures 10 & 12.

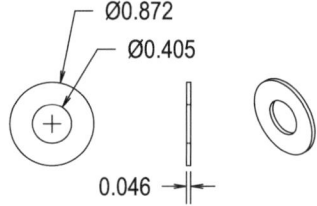

Ø0.872
Ø0.405
0.046

Figure 11. 3/8" stainless steel spacer washer detail. 19 are required.

Figure 12. The rotor assembly chucked in the lathe. The disks are aligned with a 3/8" diameter length of round rod. The alignment holes in each disk are marked with an arrow as shown in the photo.

You are now ready to reduce the diameter of the rotor disks to 2.70". Leave the alignment shaft in place for this procedure and move the tail stock forward to secure the shaft for between centers turning. Begin reducing the diameter of the turbine disks to 2.70". The disks may slip on the shaft during the process so keep an eye on the alignment shaft. If they do slip, simply realign and tighten the nut to re-secure them in position. In my experience it is best to take shallow .005" to .010" cuts while turning down the rotor disks because deeper cuts tend to grab the work and cause all kinds of aggravation. Also, this portion of the project is likely to create lots of small hot chips, so as in all metal working activity, you should be wearing eye protection.

When the disks have been reduced as stated above, take a small file and while the lathe is turning, use it to remove the burrs and sharp

Figure 13. The completed rotor assembly. The diameter has been reduced to 2.70" and the alignment tool has been removed. Note that all exhaust holes remain in alignment.

edges from each disk.

Next remove the shaft from the lathe and change to a 1/4" collet. Now chuck the 1/4" end of the shaft in the collet. Reduce the diameter of the shaft to 3/8" back 1-1/4" from the end. Next reduce the diameter to 1/4" back 1-1/8" from the end.

The shaft and rotor portion of the project is now complete and you may remove it from the lathe.

Making The Stator

The turbine stator is that portion of the turbine that remains stationary as distinguished from the rotor which is the revolving portion of the turbine. The stator is made from a 2-1/4" length of 3" O.D. .065 wall type 304 stainless steel pipe. The finished length of the stator is 2" so the extra 1/4" is added to allow for truing up the ends. See figure 14.

To prepare the stator, chuck the 3" O.D. pipe in the lathe to true up one end then turn it end for end and rechuck in the lathe to turn the other end. As stated, the finished length is to be 2". The three 1/8" holes located around the circumference of one end can be located as follows. Cut a paper strip 1/2" wide to fit around the circumference of the stator from card stock. Divide the paper strip into 3 equal divisions marking each division location 1/8" back from the edge. Wrap the paper strip around one end of the sta-

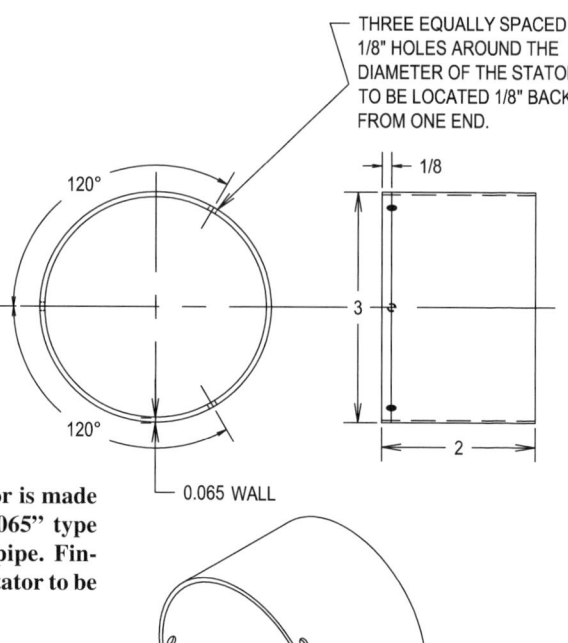

THREE EQUALLY SPACED 1/8" HOLES AROUND THE DIAMETER OF THE STATOR TO BE LOCATED 1/8" BACK FROM ONE END.

120°

120°

1/8

3

2

0.065 WALL

Figure 14. The stator is made from 3" diameter .065" type 304 stainless steel pipe. Finished length of the stator to be 2".

tor and secure its ends with tape. The edge of the paper strip should align with the end of the stator. Use the marked hole locations in the paper strip as a guide in drilling the 3 equally spaced 1/8" holes around the circumference of the stator.

The stator requires two end caps, one permanently attached by brazing or silver soldering and the other is attached with screws. The permanent end is simply a disk cut from a sheet of 20 ga. stainless steel sheet metal slightly larger than the diameter of the stator. It is then centered on one end of the stator and silver soldered in place. See figure 15.

SILVER SOLDER A 3" DIA. 20 GAGE STAINLESS STEEL DISK TO ONE END OF STATOR.

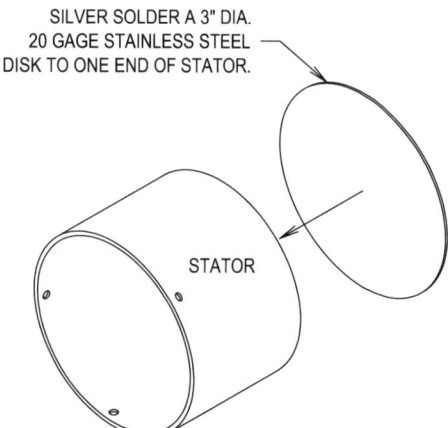

STATOR

Figure 15. Attaching the fixed end to the stator.

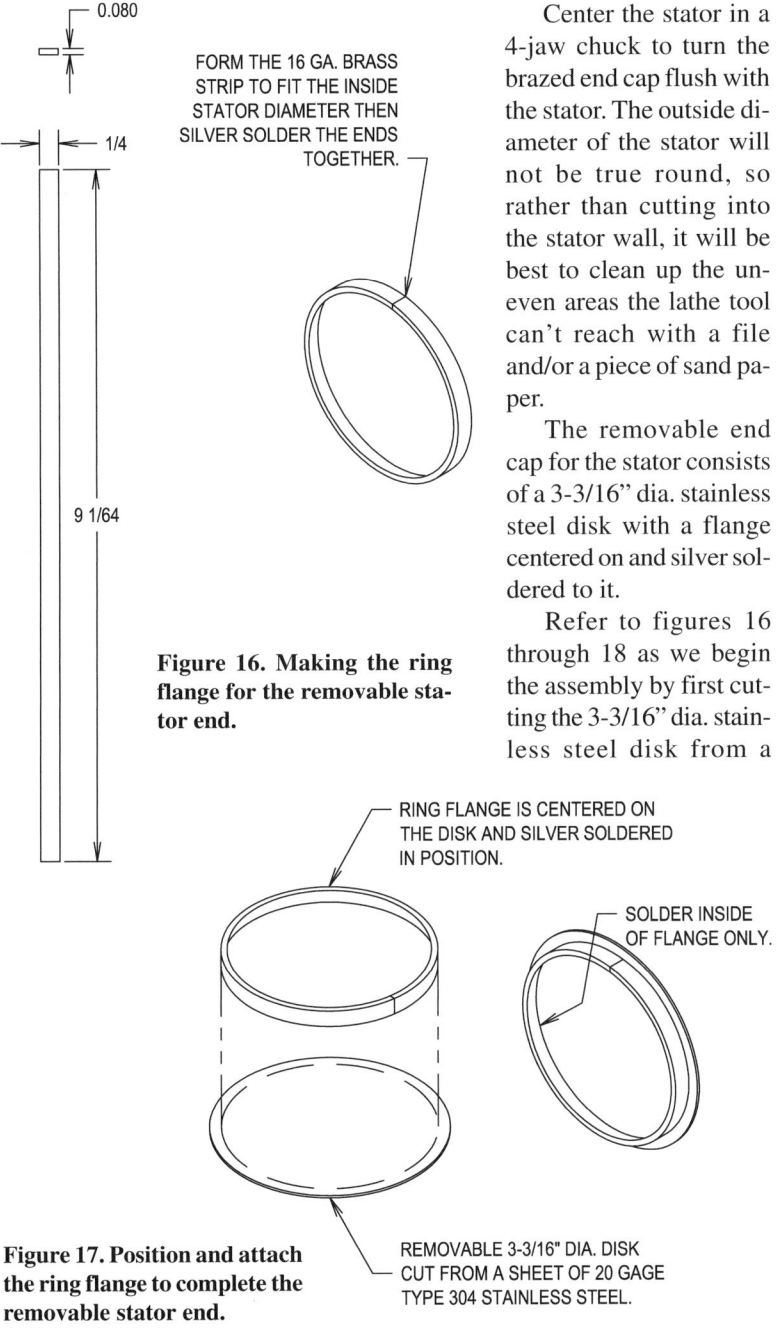

0.080

1/4

9 1/64

FORM THE 16 GA. BRASS
STRIP TO FIT THE INSIDE
STATOR DIAMETER THEN
SILVER SOLDER THE ENDS
TOGETHER.

Figure 16. Making the ring flange for the removable stator end.

Center the stator in a 4-jaw chuck to turn the brazed end cap flush with the stator. The outside diameter of the stator will not be true round, so rather than cutting into the stator wall, it will be best to clean up the uneven areas the lathe tool can't reach with a file and/or a piece of sand paper.

The removable end cap for the stator consists of a 3-3/16" dia. stainless steel disk with a flange centered on and silver soldered to it.

Refer to figures 16 through 18 as we begin the assembly by first cutting the 3-3/16" dia. stainless steel disk from a

RING FLANGE IS CENTERED ON
THE DISK AND SILVER SOLDERED
IN POSITION.

SOLDER INSIDE
OF FLANGE ONLY.

Figure 17. Position and attach the ring flange to complete the removable stator end.

REMOVABLE 3-3/16" DIA. DISK
CUT FROM A SHEET OF 20 GAGE
TYPE 304 STAINLESS STEEL.

sheet of 20 ga. type 304 stainless steel.

Cut the flange for the removable end from a sheet of 16 gage brass. It must be sized to fit the inside diameter of the stator and you can determine the required length of the brass strip using the following formula. Diameter x 3.1416= circumference. The inside diameter of the stator measures 2.87". So; 3.1416 x 2.87=9.016. 9.016" is close to 1/64" so we need a 16 gage brass strip cut 1/4" wide x 9-1/64" long. See figure 16. When cut to size, the strip is formed to fit the inside diameter of the stator. Once formed, silver solder the ends together to complete the flange.

Center the flange on the 3-3/16" diameter disk. To help center it you can scribe a circle in the center of the disk equal to the O.D. of the flange. When positioned, secure the flange on the disk with a couple of small "C" clamps and then silver solder it in place. Important! Silver solder on the inside of the flange only. If you solder the outside of the flange it will likely not fit inside the stator as it should.

When the flanged end cap has cooled, insert it in and flush against the end of the stator. Using a trans-

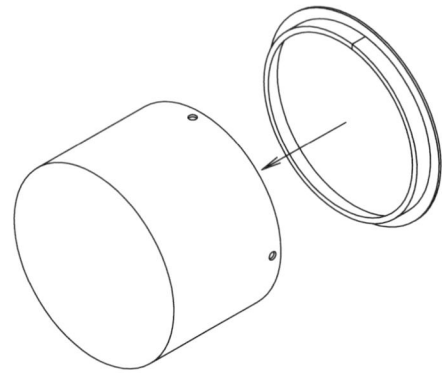

INSERT THE FLANGED REMOVABLE DISK IN THE END OF THE STATOR. USING A TRANSFER PUNCH, MARK THE LOCATIONS OF THE THREE HOLES ON THE FLANGE OF THE REMOVABLE DISK.

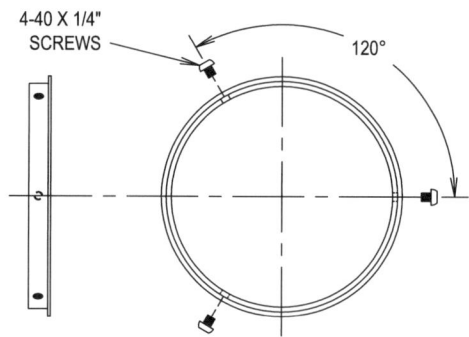

REMOVE THE FLANGED DISK FROM THE STATOR. DRILL EACH OF THE THREE HOLES 3/32" THEN TAP 4-40.

Figure 18. Installing the removable end in the stator.

fer punch, mark the hole location on the flange ring to correspond with one of the three 1/8" holes in the stator. Remove the flanged end cap and drill the marked location with a 3/32" drill then tap for 4-40 threads. Reinsert it in the end of the stator. Secure it with a 4-40 x 1/4" long screw. Then mark the other two hole locations with a transfer punch. Remove the flanged cap and drill the two holes 3/32" and tap 4-40. See figure 18.

Reinsert the flanged end and secure it with three 4-40 x 1/4" screws. Carefully center the stator in a 4-jaw chuck to true the diameter of the removable cap. Remove any sharp edges with a file. The finished diameter of the cap should be approximately 3-1/8" which allows it to extend approximately 1/16" beyond the edge of the stator wall.

For the next step leave the stator chucked in the 4-jaw. Using the tail stock chuck, step drill a 23/64" hole through the center of both ends of the stator. Then ream .376".

F.Y.I. The term step drilling refers to the process of gradually increasing the size of a hole by using progressively larger drill bits until the desired size is reached.

The Stator End Supports

The stator end supports are each made from a 3" x 3-1/2" piece of 20 gage brass sheet. A matched pair are required. Cut to size, drill, ream and tap as shown in figure 19. It is helpful to layout one end, then clamp the two ends together and drill both at the same time.

Each end support is bent to shape as shown in figure 20. Keep in mind they are an opposing pair, and it is important both end supports be formed as near alike as possible. That is, the holes line up and the height of both brackets are the same. The importance of this will

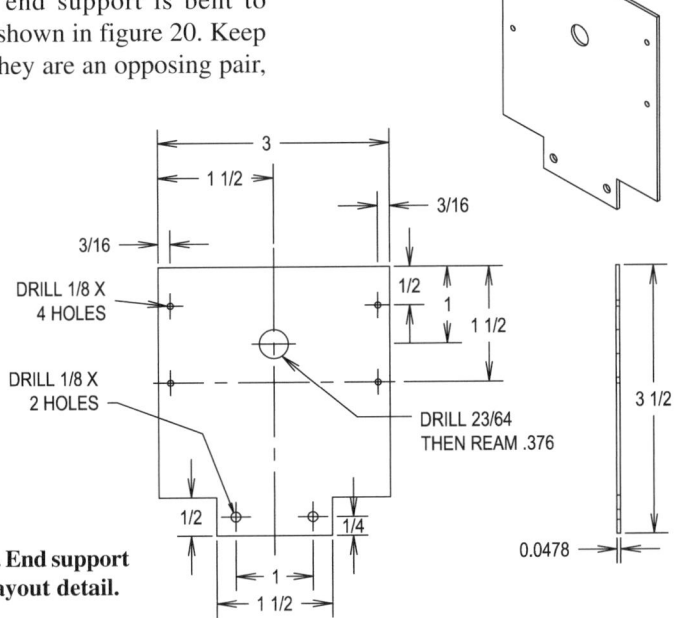

Figure 19. End support bracket layout detail.

31

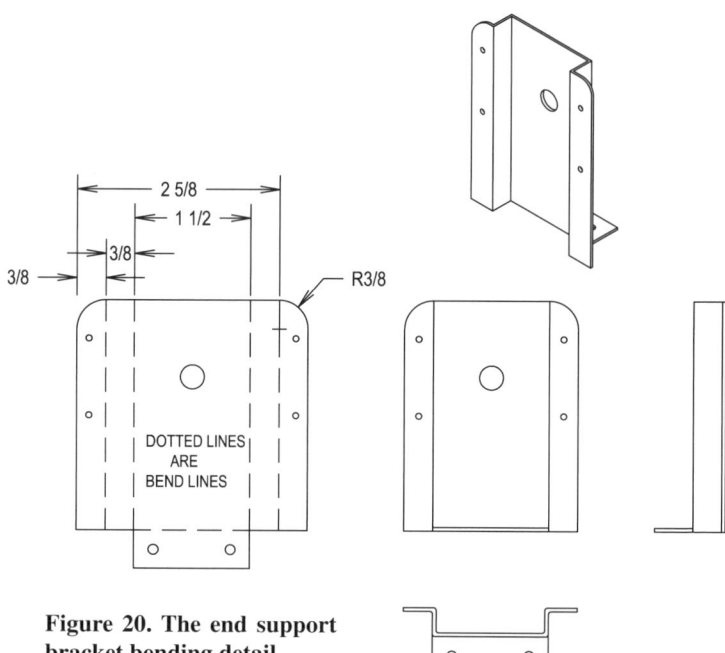

Figure 20. The end support bracket bending detail.

become more evident as you consider the next step which will be aligning and mounting the end supports to the stator. The brass is easy to form and the bends can be made with a pair of sheet metal clamp pliers in conjunction with a bench vise. Or of course bends can be made with a small sheet metal brake if you have that capability.

For the alignment procedure, you will need a 5" length of 3/8" diameter c.r.s. round rod. Insert the 3/8" rod through the .376" hole in one of the support brackets, then insert it through the center hole in the stator and finally through the .376" hole in the remaining support bracket. Refer to figure 21 for clarification.

The support brackets are fastened to the stator ends with 4-40 x 3/16" machine screws. Position the assembly on a flat surface. Ensure that one of the screws fastening the removable end of the stator is on top. Use a transfer punch to mark the locations for the holes. Rather than marking the hole locations all at once, it is best to locate one hole then remove the bracket and drill and tap it 4-40. Reposition the support bracket and secure it with the one 4-40 screw. Then locate, drill & tap each successive hole one at a time. This method will ensure the holes all align. Repeat the procedure for the second support bracket.

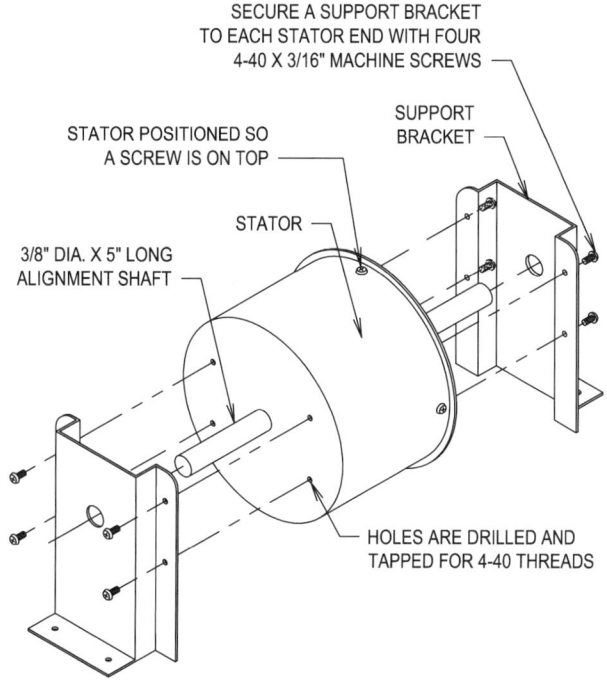

SECURE A SUPPORT BRACKET
TO EACH STATOR END WITH FOUR
4-40 X 3/16" MACHINE SCREWS

SUPPORT
BRACKET

STATOR POSITIONED SO
A SCREW IS ON TOP

STATOR

3/8" DIA. X 5" LONG
ALIGNMENT SHAFT

HOLES ARE DRILLED AND
TAPPED FOR 4-40 THREADS

Figure 21. Positioning and mounting the end support brackets.

Figure 22. Here we see the brackets mounted to the stator ends with the alignment shaft still remaining in position for the next step.

Bearings, Bearing Caps & Retainers

The bearings used in this project are classified as high speed bearings capable of running at speeds up to 30,000 rpm.

You will need two bearings. They are sold under part #'s R4A and R4AZZ. Each bearing measures .250 I.D. X .750 O.D. X .2812 wide. Mine were purchased on Ebay and I was able to buy 10 for $19.95 plus 5.95 for shipping. The Ebay store I purchased them from was 'VXB Ball Bearings'. To locate them on Ebay just type in R4A bearings on the Ebay home page search line. Or to access the VXB on line Ebay store, type VXB Ball Bearings on the Ebay store search line.

The same bearings can also be purchased on the web from ESI Bearing Distribution. Their price at the time this book goes to press was quoted at $2.79 each plus shipping. ESI's web address is http://www.bearings direct.com.

Figure 23. Bearings.

With the bearings in hand, you can begin making the bearing caps. You will need two, and they are made from 1" diameter brass round rod. See figure 24 for dimensional details. Begin making the bearing caps by chucking a 3" length of 1" diameter brass round rod in a 3-jaw lathe chuck. Face off the end with the lathe tool. Center drill the end and with the tail stock chuck, drill 2" deep using a 23/64" drill. Ream the 23/64" hole to .376" using a chucking reamer. Reduce the outside diameter to .875", .295" back from the end. Part off at .375" long.

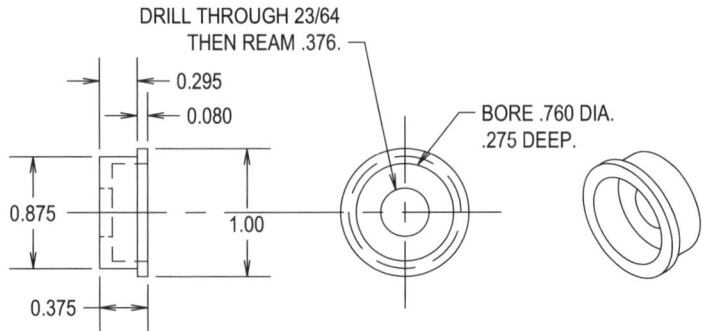

Figure 24. Bearing cap. Make two from 1" diameter brass round rod.

Then repeat the above procedure to make the second bearing cap.

To complete the bearing caps, chuck them in the 3-jaw from the .875" end. Bore the inside diameter to .760", .275" deep.

The bearing retainers shown in figure 25 are made from a piece of 16 gage brass sheet. Two are required. Scribe a circle with a set of dividers to locate the 1-3/8" finished diameter. Then cut out both to rough size with a bandsaw or hacksaw. Use the dividers once again to scribe the 1-1/8" diameter circle in the center of each retainer. Drop a vertical center line to locate the position of the 7/64" holes. Then drill the two 7/64" holes in each retainer blank as shown in the figure.

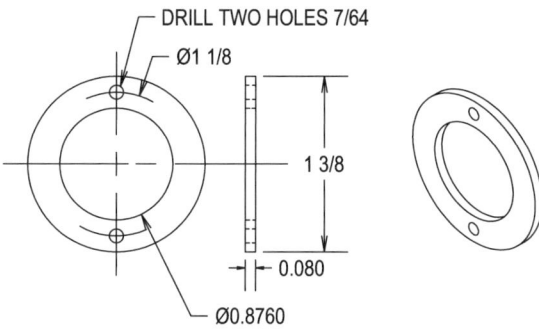

Figure 25. Bearing retainer detail. Make two from a piece of 16 gage brass.

The arbor required for turning the finished O.D. of the retainer rings can be made from a 3" length of 1" diameter c.r.s. round rod with a 1/4-20 hole drilled and taped in one end. The arbor is mounted in the 3-jaw lathe chuck. To prepare the retainer ring for mounting on the arbor, drill a 1/4" hole through its center. Then use a 1/4-20 bolt and flat washer to attach it to the arbor. Reduce the diameter of the retainer ring to 1-3/8". See figure 26 for the set up.

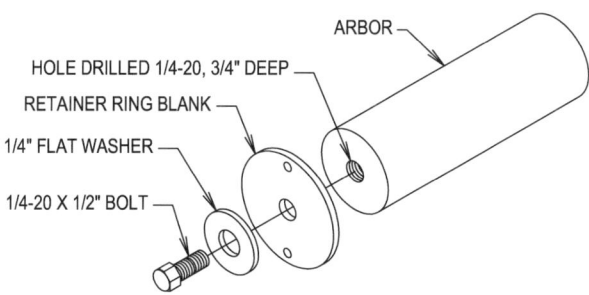

Figure 26. Arbor used for turning the outside diameter of the bearing retainers. Make one from a 3" length of 1" diameter c.r.s. round rod.

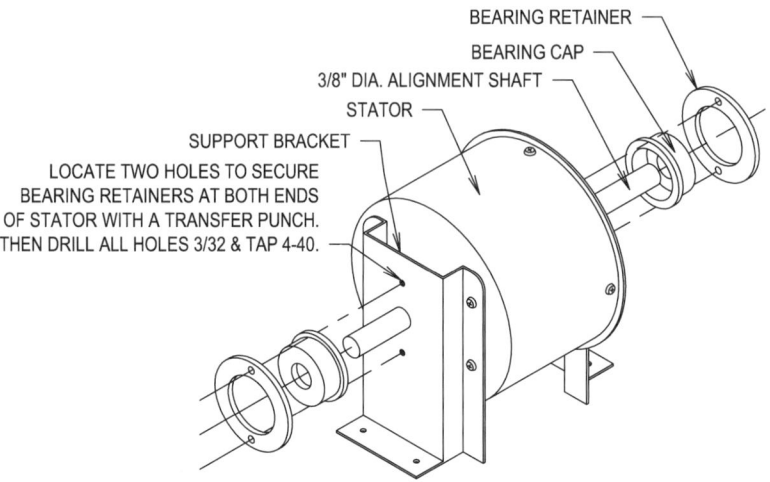

BEARING RETAINER

BEARING CAP

3/8" DIA. ALIGNMENT SHAFT

STATOR

SUPPORT BRACKET

LOCATE TWO HOLES TO SECURE
BEARING RETAINERS AT BOTH ENDS
OF STATOR WITH A TRANSFER PUNCH.
THEN DRILL ALL HOLES 3/32 & TAP 4-40.

Figure 27. The drawing shows the procedure for locating the retainer ring mounting holes and securing the bearing caps.

Remove the retainer from the arbor and center it in the lathe chuck to bore the .876" I.D..

The bearing cap and retainers can now be positioned on each end of the turbine. Slide a bearing cap on the 3/8" diameter alignment shaft followed by a retainer ring. Align the 7/64" holes in the retainer ring on vertical center, and using a transfer punch, mark the location of the two holes in the retainer ring on the support bracket. Then drill and tap the marked holes in the support bracket for 4-40 threads. Repeat the procedure for the opposite side of the stator. Secure both bearing caps and retainer rings to the brackets with 4-40 x 3/8" machine screws. See figure 27 for an illustration.

Remove the alignment shaft and the support brackets from both ends of the stator. Also remove the bear-ing retainers and caps from both support brackets. It's a good idea to reference each bracket and bearing retainer with a mark to ensure each goes back on the correct end during reassembly.

Figure 28. Photo shows the bearing cap and retainer ring mounted in position.

36

With both support brackets removed, chuck the stator in the lathe securing it from its fixed end, and using a boring bar, increase the I.D. through both ends to 1.375". See figure 29.

The rotor & stator are ready for reassembly. First install the bearings in the bearing caps, and using the bearing retainers, secure them to the support brackets. Remount the support brackets on each end of the stator.

To install the rotor you will have to once again remove the end from the stator. This time it is OK to leave

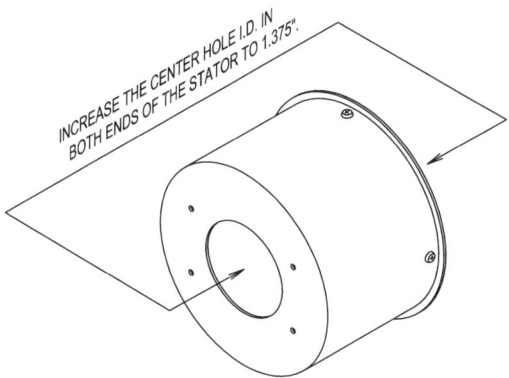

Figure 29. Increase the center hole I.D. in both ends of stator to 1.375".

the support bracket attached to it.

Before inserting the rotor in the stator, you will need to make a 1/4" I.D. spacer bushing for each end. The spacer detail is given in figure

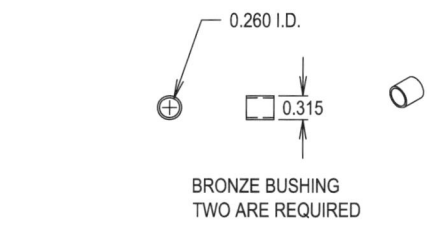

BRONZE BUSHING
TWO ARE REQUIRED

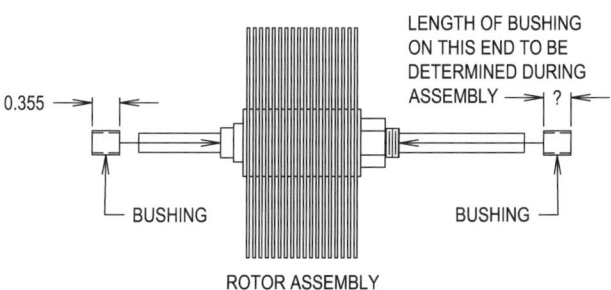

Figure 30. Make & install bronze bushings for each end of the rotor shaft.

30 and were satisfactory for my turbine. But slight length variations in these dimensions may exist from turbine to turbine. What is required is that the rotor be centered in the stator as near as possible and that it have no side play.

With the spacer bushings made, the rotor can be installed in the stator as shown in figure 31. Note that the rotor end with the nut securing the disks faces the direction of removable end. This is also the end with longest shaft. The extra shaft length on this end allows for a pulley if so desired. With a pulley you may be able use the turbine to power a small alternator or generator.

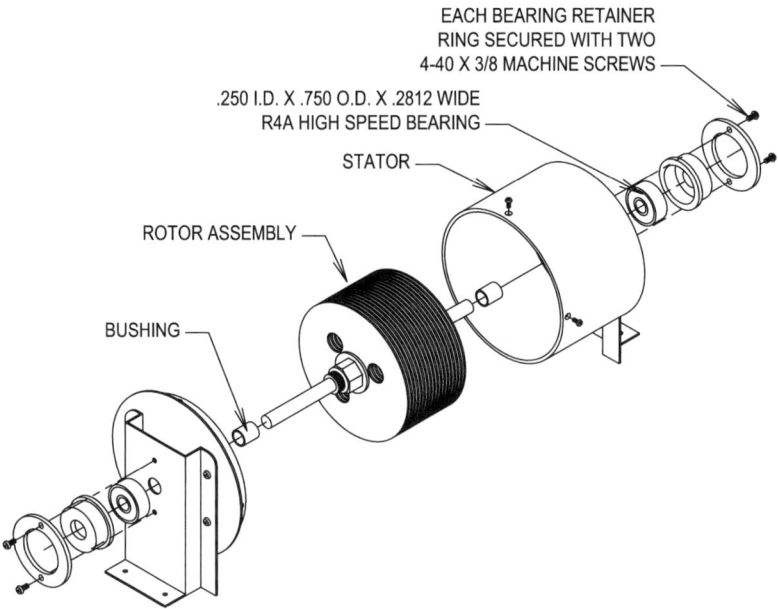

EACH BEARING RETAINER
RING SECURED WITH TWO
4-40 X 3/8 MACHINE SCREWS

.250 I.D. X .750 O.D. X .2812 WIDE
R4A HIGH SPEED BEARING

STATOR

ROTOR ASSEMBLY

BUSHING

Figure 31. The turbine assembly.

Inlet Plumbing & Nozzle Detail

With the turbine complete, construction can begin on the pressure inlet assembly. But first make a base for the turbine from a 3-1/2" x 6" block of 3/4" thick wood as shown in figure 32. I used red oak, but the preference is yours. Two 5/32" holes are drilled in one end of the block and then counter drilled to 1/2" diameter 1/4" deep. These holes will be used for attaching the manifold. Also, it's not necessary now, but later a decorative routered edge can be applied to the mounting block along with some stain and a couple of coats of urethane.

The manifold is made from a 3-5/8" length of 1" x 1" aluminum bar stock. See figure 33 for dimensional details. Center the manifold blank in a 4-jaw chuck and face off both ends. Finished length to be 3-1/2". Center drill, then drill the main supply line hole 3-5/16" deep using a 21/64" drill. Then tap the end of the main line hole for 1/8-27 pipe thread 1/2" deep.

Remove the manifold from the lathe chuck. Locate and mark the position of the two holes on the top of the manifold and the two holes

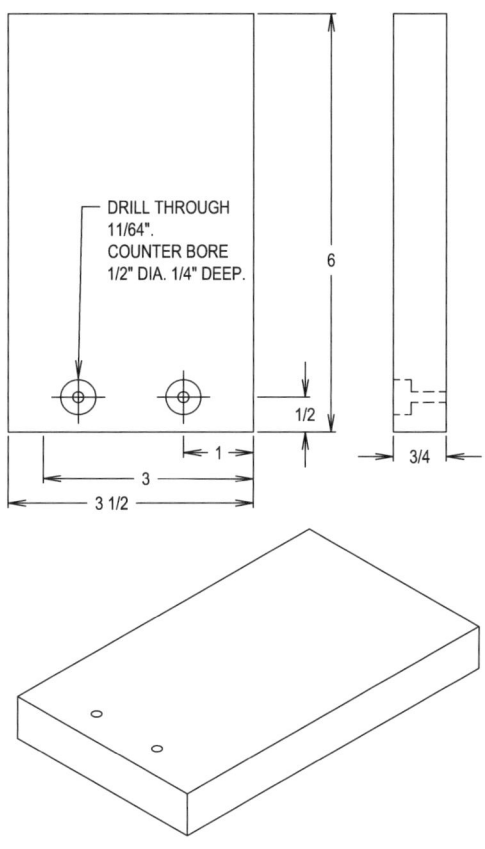

Figure 32. Mounting block.

on the bottom of the manifold. Drill the top two holes 21/64" through to the main supply line hole and tap for 1/8-27 N.P.T. threads. Drill the two bottom holes in the manifold with a #29 drill through to the main supply line hole and tap for 8-32 threads.

The manifold is now complete and it can be mounted to the base

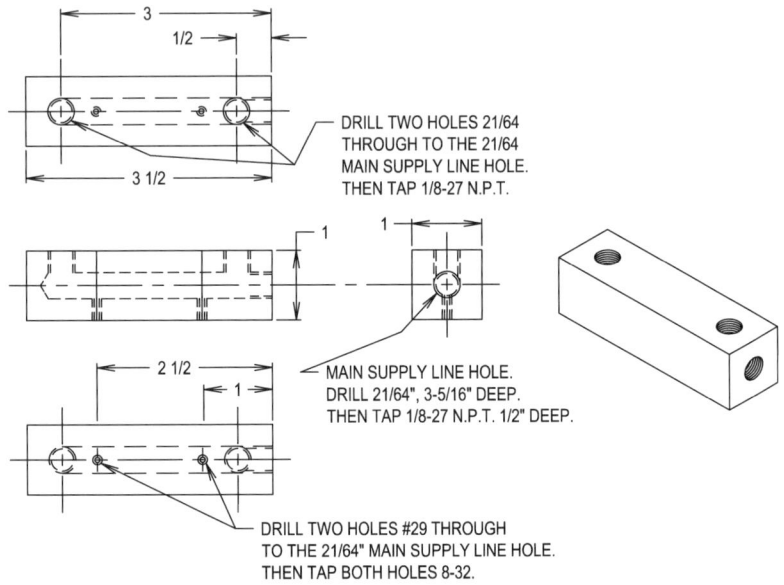

DRILL TWO HOLES 21/64
THROUGH TO THE 21/64
MAIN SUPPLY LINE HOLE.
THEN TAP 1/8-27 N.P.T.

MAIN SUPPLY LINE HOLE.
DRILL 21/64", 3-5/16" DEEP.
THEN TAP 1/8-27 N.P.T. 1/2" DEEP.

DRILL TWO HOLES #29 THROUGH
TO THE 21/64" MAIN SUPPLY LINE HOLE.
THEN TAP BOTH HOLES 8-32.

Figure 33. Manifold. Make 1 from a 3-5/8" length of 1" x 1" aluminum bar.

with two 8-32 x 1" machine screws and & two #10 flat washers as shown in figure 34.

We are now ready to make the nozzle or nozzles and assemble the pressure lines for the turbine.

Here you may choose to construct either a single nozzle turbine or a double nozzle turbine. If you build a single nozzle turbine the direction (either clockwise or counter clockwise) in which the turbine will rotate will be determined by which side of the stator you install the nozzle on. Refer back to figure 2 on page 5 for clarification.

You can also choose to install nozzles on both sides of the stator in which case you will be able change the direction of rotation by controlling the pressure flow to the desired side of the turbine with shut off valves.

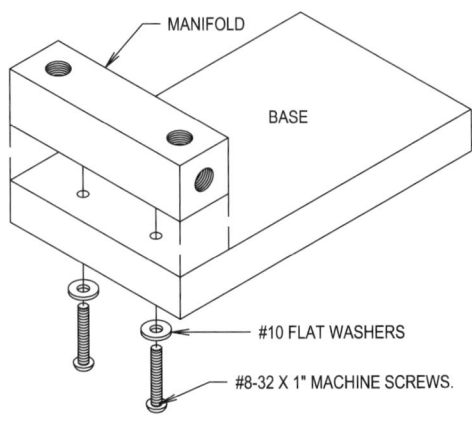

MANIFOLD

BASE

#10 FLAT WASHERS

#8-32 X 1" MACHINE SCREWS.

Figure 34. Mounting the manifold on the base.

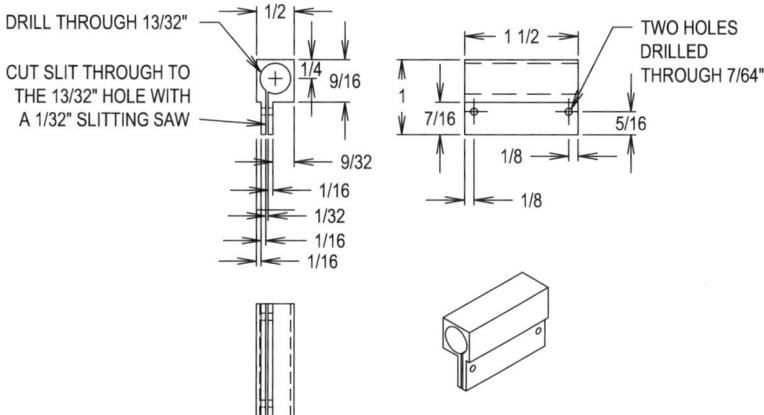

Figure 35. Nozzle. Make two from 1/2" x 1" x 1-5/8" aluminum bar.

With the that in mind, we are ready to make the nozzle or nozzles depending on your preference.

Each nozzle is made from a 1-5/8" length of 1/2" x 1" aluminum bar stock that has been faced off on both ends to a finished length of 1-1/2". See figure 35 for details.

The first step in making the nozzle is to drill the 13/32" through hole. That's best done in a lathe using the 4-jaw chuck. The cutouts in the bar stock to form the nozzle end can be made by clamping the bar stock in the mill vise. Using a slitting saw chucked in the milling machine you can make the cuts to form the nozzle énd. I used a 2-1/2" x 1/32" slitting saw to make all of the cuts to form the nozzle end. Once the nozzle is formed, locate and drill the two 7/64" holes in each end of the nozzle as shown in the detail.

Shims are needed to keep the nozzle open and prevent it from closing as the machine screws are tightened to secure the nozzle on the 4" pipe nipple. The shims can be cut from .015 thick paper card stock. My shims were cut from a manila file folder which turned out to be about the right thickness. See figure 36 for the shim detail. The holes shown in the shims were punched with a scratch awl.

Insert a shim in each end of the nozzle aligning the 3/32" hole in the shim with the 3/32" hole in the nozzle. Insert a 4-40 x 1/4 machine screw through the 7/64" hole and secure with a 4-40 nut. Do not tighten the screws at this time.

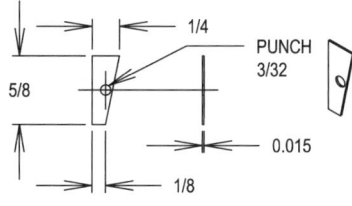

Figure 36. Nozzle shim. Make 4 from .015 thick paper card stock.

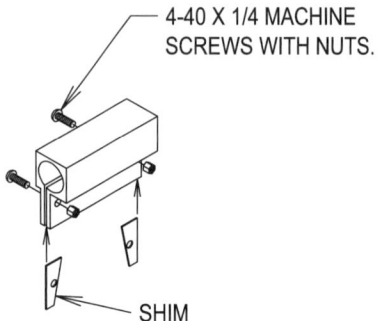

4-40 X 1/4 MACHINE
SCREWS WITH NUTS.

SHIM

Figure 37. Installing the shims.

The pressure line assembly details are given in figure 38. Assemble the individual parts as shown, but do not tighten the screws on the nozzles at this time. To prevent air leaks, seal each thread joint with plumbers tape or pipe thread putty.

Before the pressurized air can enter the turbine, slots must be cut

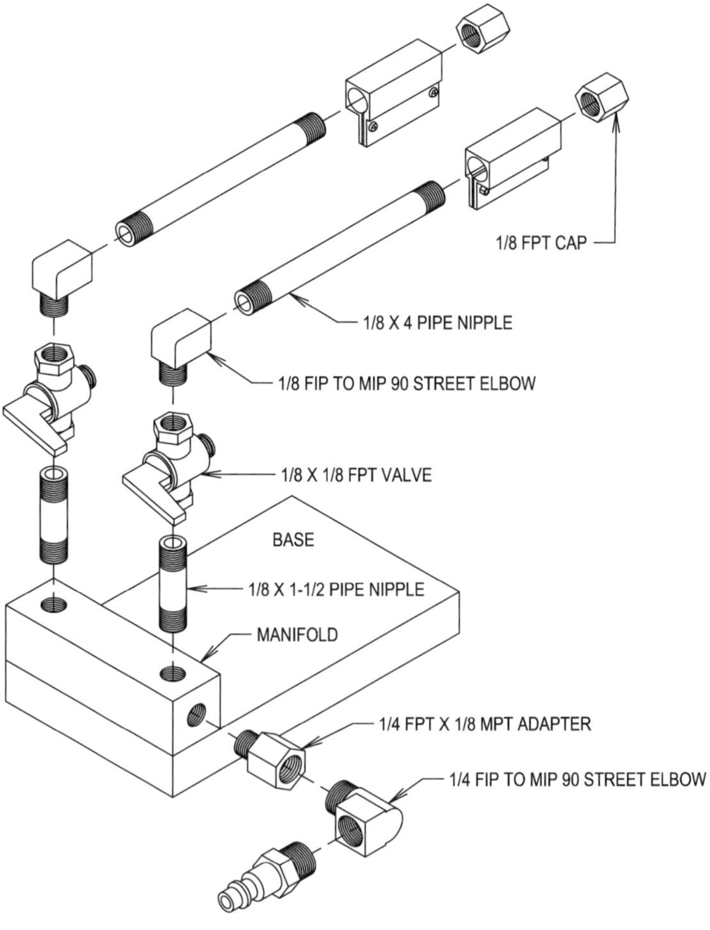

1/8 FPT CAP

1/8 X 4 PIPE NIPPLE

1/8 FIP TO MIP 90 STREET ELBOW

1/8 X 1/8 FPT VALVE

BASE

1/8 X 1-1/2 PIPE NIPPLE

MANIFOLD

1/4 FPT X 1/8 MPT ADAPTER

1/4 FIP TO MIP 90 STREET ELBOW

Figure 38. Pressure line assembly detail.

in the top of the stator for the nozzle end entry points. The slots need to be 3/16" wide x 1-1/2" long and located at approximately the 10 o'clock and 2 clock position and centered front to back.

Figure 39 shows how to find the slot locations. As the detail indicates, the turbine is centered on the base and moved forward until flush with the nozzle ends. As is indicated in the drawing, the nozzle ends are pointing straight down. Then a scribed line is extended from the in-

side corner of each nozzle end across the top surface of the turbine. Mark off the 1-1/2" length of the required slot so that it is centered on the turbine from front to back. See figure 40.

Before milling the slot in the stator, the turbine must be completely disassembled. To prepare for milling the slots in the stator wall, you will need to chuck a 3/16" drill bit in the chuck. Then clamp the stator in the mill vise positioning it so that one of the slot locations is cen-

CENTER THE TURBINE ON THE BASE BOARD AND FLUSH AGAINST THE NOZZLE ENDS. THEN EXTEND A SCRIBED LINE FROM THE INSIDE EDGE OF EACH NOZZLE ACROSS THE TOP OF THE TURBINE. THE SCRIBED LINE WILL BE USED AS A REFERENCE LINE WHEN MILLING THE 1/8" WIDE SLOTS IN THE TOP OF THE TURBINE WALL FOR THE NOZZLE ENTRY.

SCRIBED LINES

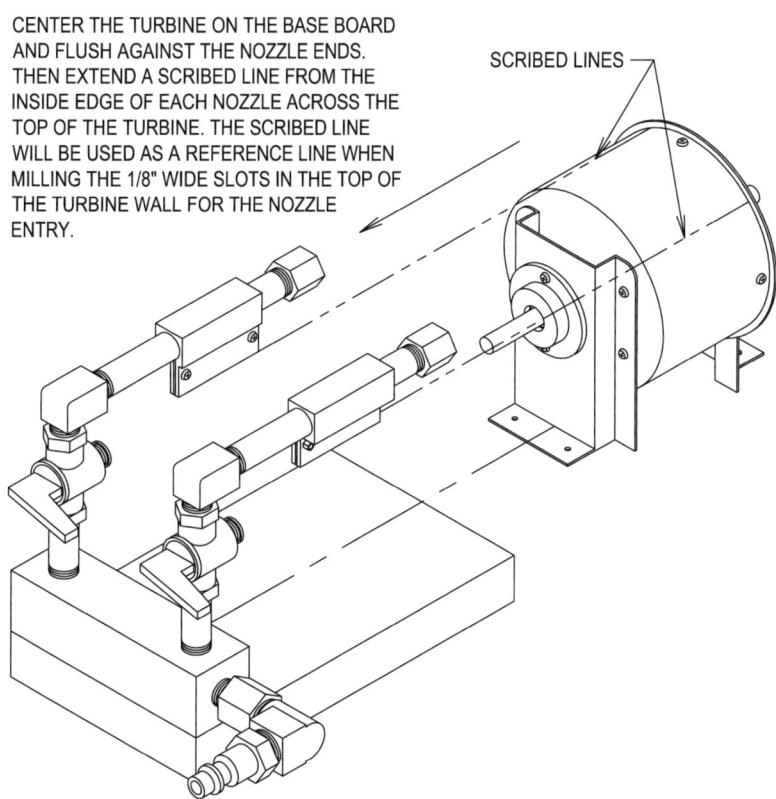

Figure 39. Locating the nozzle entry slots in the stator.

tered and perpendicular to the drill chuck. Drill a starter hole as shown in figure 41. Repeat the procedure to drill a starter hole in the other slot. Replace the drill bit with a 3/16" end mill. Now reposition the stator so the slot locations are at the 10 o'clock and 2 o'clock position in reference to the mill chuck. See figure 42. Using the 3/16" mill, cut a slot as shown in the photo. Adjust the mill table to cut the second slot.

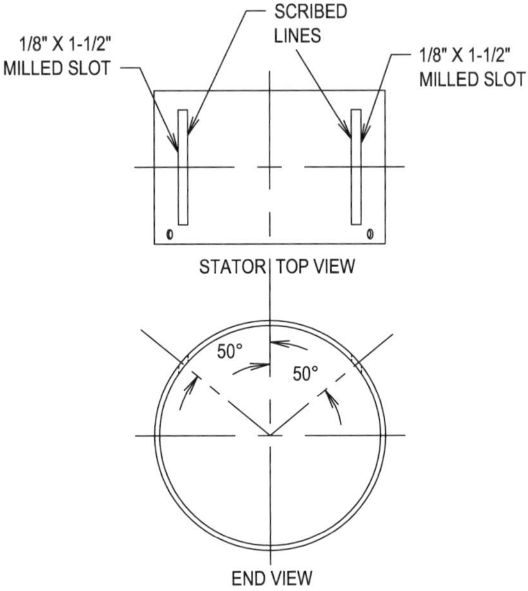

Figure 40. Detail showing the approximate locations for the slot cutouts in the stator wall.

After the slots have been milled, remove any sharp edges and reassemble the turbine.

Center the turbine on the mounting block placing it flush against the manifold. It may be necessary to swing the nozzles up to one side as the turbine is positioned. When in position, secure the turbine to the base with four #6 x 3/4" brass wood screws. See figure 43. Then slide the nozzles in position to align with the slots in the stator.

The 4" pipe nipples are prepared next. 1-1/4" long x 1/8" wide slots must be milled in the bottom of each one. The slots are located corresponding to the outlet location of the nozzle ends. Use the procedure as shown in figure 44 which shows

scribing lines to locate the slots. Then unscrew the nipples from the pressure line assembly and remove the nozzles from them as well. Note

Figure 41. Drilling a 3/16" starter hole in the stator wall.

44

Figure 42. Milling one of the 3/16" slots in the stator wall.

centered between the scribe marks made earlier. Then clamp each nipple in the mill vise to mill the required slot with a 1/8" end mill.

Re-thread the 4" long nipples into their respective elbows. Slide the nozzles back in position aligning them with the slot cut out in the nipple and the entry slot in the stator wall. When correctly positioned, tighten the screws to clamp the nozzles securely to the pipe nipples. With both nozzles positioned and secured, thread the end caps on and tighten them securely.

The nozzle ends should extend just inside the stator wall, but ensure that they do not rub against the rotor as it turns. It may be neces-

in figure 45 that the actual length of the nozzles are 1-1/2", but the length of the slot in the nipple is to be only 1-1/4". So mark a 1-1/4" long slot location in each nipple

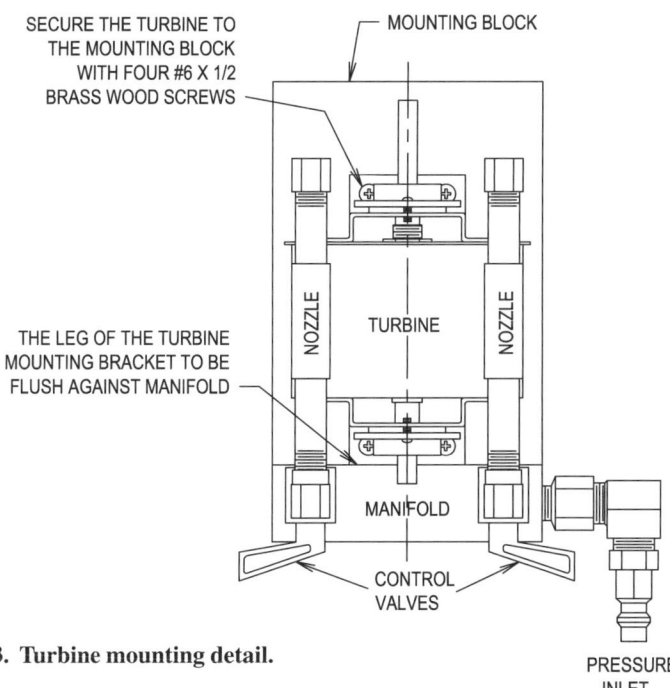

SECURE THE TURBINE TO THE MOUNTING BLOCK WITH FOUR #6 X 1/2 BRASS WOOD SCREWS

MOUNTING BLOCK

THE LEG OF THE TURBINE MOUNTING BRACKET TO BE FLUSH AGAINST MANIFOLD

NOZZLE

TURBINE

NOZZLE

MANIFOLD

CONTROL VALVES

PRESSURE INLET

Figure 43. Turbine mounting detail.

sary to cut shims and place them between the mounting legs of the support brackets and the wooden base to raise the turbine closer to the nozzles. The last turbine I built required 1/8" shims. The reason for the variation is that the pipe fittings do not always tighten to the same depth.

Congratulations! The turbine is now complete and ready for a test run. Connect the air supply, open the desired valve depending on your direction preference while leaving the opposite valve closed. Apply air pressure and a way she goes.

Operation of the unit is simple and straight forward. An air hose is connected to the manifold. Air pressure is applied and the speed of the turbine is controlled by increasing or decreasing that air pressure. And the rotational direction is controlled by opening and closing valve A & B as shown in figure 2 on page 5.

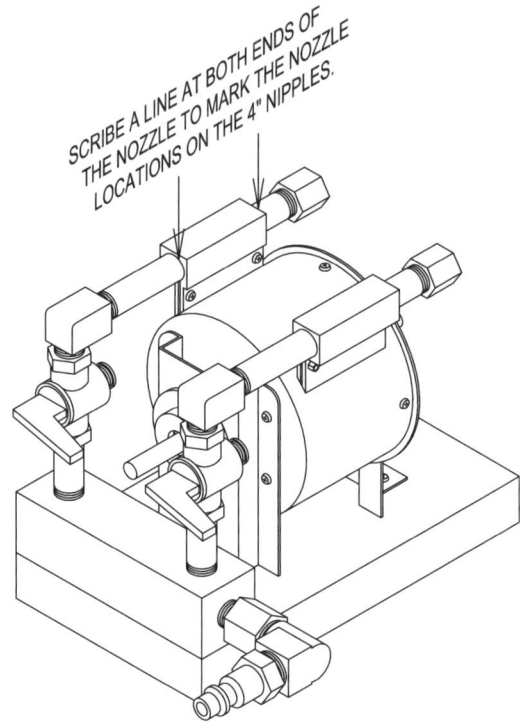

Figure 44. Marking the nozzle locations on the 4" nipples.

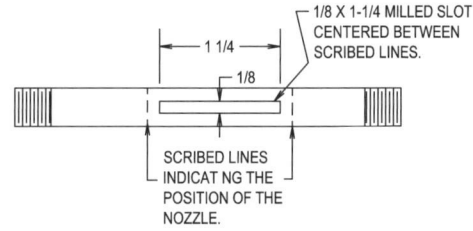

Figure 45. Slot detail for the 4" nipples.

Conclusion

Congratulations once again on the completion of your Tesla turbine. There are a few things to consider as we wind this project up.

First of all, you will likely experience air leaks between the nozzle and the 4" nipple as indicated in figure 46. When I had initially began construction of the turbine I had considered making the nozzles out of brass and then silver soldering them to the 4" nipples. I had also considered silver soldering the nozzle ends in their opening on the stator wall. And that is certainly a viable, but permanent option you might want to consider.

However, I was able to remedy the leaks with silicon gasket sealing compound. I found that a Permatex product called ULTRA GREY RTV silicone works very well for this purpose. It's gray color blends in well too. The sealing procedure is to remove the nozzle screws and shims. Spread a small amount of silicon on each of the shims and then reinstall them. Reinsert the screws and slide the nozzles back on the 4" nipples. Apply a little silicon on the 4" nipple near each end of the nozzle. Reposition the nozzle and tighten the screws to clamp them in place. You will want to ensure that the silicon does not block the nozzle ends. Allow the required time for the silicon to set, usually 24 hours and your ready to go again.

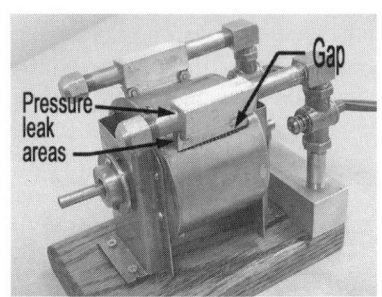

Figure 46. Photo showing potential air leak areas and the gap that exits where the nozzle enters the turbine. See text.

Also in figure 46, notice where the arrow points to the gap between the nozzle end and the point where it enters the stator wall. The turbine will operate fine with these gaps. Essentially there will be a venturi effect on the inlet nozzle. And there will be some blow back on the opposite or closed side nozzle. There is a positive aspect to these gaps in that it offers an opportunity to show interested ones a glimpse of the flat disk rotor inside without having the disassemble the unit. However the unit will run quieter and you will have more efficient use of air pressure if the gaps are sealed. If you decide to seal these gaps you can do so with the same gray rtv silicon. If care is taken to run a small, neat and tidy bead the job will look very professional. However, be extremely careful not to allow silicon to cover over or get into the nozzle ends. A suggestion would be to remove the front end of the stator. You can leave the front mounting bracket attached to the removable end, but you will need to remove the two screws that secure it to the base. Of course to remove the front end from the stator you simply remove the 3 screws securing it and then pull it away. Then remove the rotor assembly which will allow you to reach inside the stator to tape over each nozzle end before applying the silicon. When you're done simply remove the tape. Also, if the gaps around the nozzle ends are very wide you may want to lay a tape backing to close the gap. This should be done from inside the stator as well which will prevent the silicon from falling through the gaps.

When the silicon has cured reassemble the turbine and your ready to go again.

The appearance of the turbine can be improved by rounding the edges of the base with a router. After sanding down the base, you might want to consider applying stain and urethane. Also, other parts of the turbine such as the stator can be painted if desired.

One of the fascinating things about this project is that it has so much room for experimentation. Even this little model turbine has potential in that it could be used to power a small generator or alternator. Of course using air pressure for other than demonstration purposes is not very practical from an economical standpoint, but steam pressure would be. In fact that's the way many power plants operate today. Coal fired boilers produce steam that powers turbines very similar, but much larger than what you have constructed from this manual. Be advised though; steam boilers are extremely dangerous! You should only venture down that road if you have experience in such matters and then only after careful study and consideration.

Other experiments can be conducted with converting the turbine into a water pump, air pump or even a vacuum pump. W.M.J. Cairns discusses these and other possibilities in his book titled "The Tesla Turbine'. If you want to go further with the Tesla concept, I would encourage you to read Cairns' book.

Best wishes for success in this and all of your future shop projects.

Vince Gingery